INTRODUCING TARANTULAS

The attitude of the average person to spiders is usually one of ambivalence or hatred. While those of the first group are aware that spiders catch flies, which is beneficial to the average householder, such people dislike the cobwebs they build for this purpose. Further, the sight of a rather fat spider running across their carpet or kitchen floor creates a shudder of fear for many people. Such fear is derived from their childhood days when their mother would be seen to stamp on anything that scurried across the floor. The fear was then mentally reinforced by fairy tales and folklore in which spiders were portrayed as creatures that could crawl over you and bite.

Then there are the horror movies in which giant mutated spiders run amok,

become quite popular. From an educational viewpoint this is wholly for the good, and a number of schools are now keeping these (and other) spiders as projects of study.

Given that spiders are so common it is perhaps surprising that they are generally so misunderstood—or is it because they are so readily seen that people simply do not take the trouble to find out a little about them and their lifestyle? If they did, far fewer people would enthusiastically go about the business of stamping on them and in other ways trying to reduce the population of them in their homes.

The fact is that contrary to popular belief only a handful of the thousands of

A Chilean Beautiful Tarantula, *Grammostola cala*, one of the more commonly seen pet tarantulas. Make a good choice for your first tarantula and you will have a wonderful pet for years to come. Photo: M. Gilroy.

species represent any form of threat to humans, and even these do so only if their private domain is invaded, be this inadvertently or deliberately. All spiders are poisonous, but only a handful have the power to actually inject their venom into your skin. Of those few that can, only an even smaller number can do any more than cause local pain rather in the manner of a bee or wasp sting. Indeed, the sting of the latter is generally far more painful than that of a spider, with but few exceptions.

destroying entire buildings and munching their way through a segment of the human population, totally immune to guns, explosives, and fire! If they are not doing this, then an overgrown hairy specimen is crawling over James Bond or falling on the arm of a terrified female. The said hairy beast was of course a tarantula, so this little furry animal hardly seems a likely candidate as a pet.

But these spiders have always attracted the interest of a number of people. Today, the so-called tarantulas have

The popular pet tarantulas are not among those exceptions and many are in fact among the most docile of spiders as far as humans are concerned, even though they have a formidable appearance. For this reason they can be safely kept in your home without the fear that if they get loose your very existence will be in peril!

This does not mean that they should be handled with impunity, as you will discover. But you can be assured that if you care for your tarantula as you should the chances of it ever biting or otherwise inflicting pain on you is lower than the risk of a pet hamster or mouse biting you or a cat scratching you. These and other pets can actually inflict far more damage than can the tarantula or most other spiders.

In the following chapters you will find all of the information you are likely to need in order to both appreciate and care for these fascinating creatures. From an understanding of their natural history to breeding them, the text makes no assumption of previous knowledge. It will provide you with a sound insight not only to tarantulas but to spiders in general.

It often is difficult to identify tarantulas to species or even genus. Depend on your pet shop to steer you to a calm, long-lived species. This is probably a South American Pink-toe, *Avicularia*.

M. GILROY

HISTORY OF TARANTULAS

Spiders are members of a group (scientifically called a phylum) of animals known as Arthropoda. This phylum is arguably the most successful group of animals on Earth, far more so than the phylum Chordata (which includes humans, other mammals, birds, reptiles, amphibians, and fishes), at least in terms of numbers of species. When the last human eventually vanishes from Earth you can be sure there will still be thou-

Spiders have been around for a long time, as shown by this fossil from the Cretaceous of northeastern Brazil.

D. A. GRIMALDI, AMNH

sands of arthropod species doing very nicely. These animals are very diverse in both their form and habitat. Some members live in the frozen wastes, some in the depths of the deepest caves. Some are quite happy living in scorching deserts, while others inhabit the seas and freshwaters of our planet. Some are terrestrial, others can fly.

The term arthropod means jointed leg. It applies to such animals as insects, millipedes, centipedes, crabs, lobsters, crayfish, scorpions, ticks, and the spiders. There are about one million identified arthropod species, which account for about 85% of all known living organisms, but it is estimated that there remain perhaps another million or more species yet to be identified! The most successful group of these animals is the

insects, which number at least 850,000 species.

However, while not comparable in number of species with the insects, the spiders are very successful when compared with other animal groups. There are about 30,000 species of bony fish, 6,000 reptilians, 3,000 amphibians, 9,000 bird species, and 4,500 mammals, but there are around 35,000 species of spiders—and this is thought to represent only a quarter of the number that actually occur. If this is so, then there remain over 100,000 species still to be identified!

ANATOMY

Arthropods do not have an internal skeleton like mammals, fishes, reptiles, or birds. Instead they have a stiff outer covering (exoskeleton) made up of layers of carbohydrates and protein, much of which is a chemical called chitin. The outer layer is waterproof. The muscles of the body and appendages are fastened to the underside of this exoskeleton, which is flexible. The skeleton is thus akin to a suit of armor that gives good protection to its owner. However, because this "armor" is so hard it does not grow. Arthropods must therefore go through a series of molts until they reach adult size and in some instances, including crustaceans and certain spiders, after maturity has been reached. At such times they are vulnerable to being attacked by other creatures and they must stay in cool places otherwise they may become too dry, negatively affecting the molting process.

The body of arthropods is segmented and is typically of two or three distinct parts. The heart is tube-like and lies under the dorsal surface of the abdomen. Blood flows through arteries and veins but it openly bathes the body cells rather

M. GILROY

This head-on view of a Chilean Beautiful shows not only the two major divisions of the body, but the large fangs or chelicerae and the pedipalps as well.

than feeding them via capillaries (an open system). If such an animal should have its body pierced, the consequence of the system is that it loses blood much more rapidly than does an animal possessing a capillary or closed system. The nervous system is found on the ventral side of the animal. The number of legs arthropods may have ranges from six to several hundred, though spiders have only eight.

THE ARACHNIDS

There are about 75,000 described species of arachnids (half of them mites). They differ most obviously from the insects in that they lack antennae and have four pairs of legs rather than three. They also have only two external body divisions. Insects typically have three divisions—the head, thorax, and abdo-

men—but in arachnids the head and thorax are fused to form what is called the cephalothorax. It is to the cephalothorax that the legs of arachnids, thus spiders, are attached. The abdomen contains most of the major bodily organs other than the brain and muscles for the legs and mouthparts.

The "head" of arachnids contains as many as 12 simple eyes as well as two other pairs of appendages. One pair is the fangs or chelicerae, which have poison glands at their bases. The other paired appendages are a pair of pedipalps that look like small legs and are used as an aid in handling food, transferring sperm from the male to the female, and as sensory organs. The eyesight of arachnids is not especially good for most species, but serves to detect levels of light and

A tarantula's eyes are placed in a cluster or turret near the middle front of the cephalothorax. Tarantulas rely on their special bristles or setae to detect movements, not their eyes. Photo: M. Gilroy.

dark. Only in a limited number of species are the eyes capable of forming distinct images.

In size, arachnids range from tiny mites that parasitize many animals, to a large African scorpion that may exceed 18cm (7in) in length. The arachnids, unlike the insects, feature neither antennae nor wings. They are divided into about five major groups, known as orders, the spiders being in order Araneae.

THE SPIDERS

Having looked briefly at the main features of arthropods and arachnids, we can now focus more specifically on the spiders. These range in size from a tiny species measuring about 0.4 cm (0.16 in) that is found on the island of Samoa to the giants of the spider world that may attain a body length of over 8.75 cm (3.5

R. BECHTER

This rare tarantula, *Megaphobema robusta*, represents just one of the many surprises that you could find in your local pet shop. Remember that most tarantulas are very similar and require similar husbandry, with some exceptions.

in) and a total leg span of maybe 28 cm (11 in). The latter are the "bird-eating" spiders of tropical forests. It is the spiders of this latter type that have become known as tarantulas, though most do not of course attain the size just cited.

Members of the order Araneae are divided into three major groups or suborders, but all the tarantulas belong to the suborder Orthognatha. This is the group of spiders we are most interested in, and they are often collectively called the mygalomorphs. They are placed into 11 families, within which are to be found the common tarantulas, the gigantic bird-eating spiders, and the much feared funnelweb spiders of Australia, as well as the many tarantula-like spiders found in other parts of the world.

Many orthognathids live in burrows. These may have a trapdoor entrance or a silken web that is fashioned rather like a funnel, thus their common name of funnelweb spiders. Others build a web that resembles a tube or purse, and these are known as the purseweb spiders. Our interest lies mostly with certain hairy species that burrow into the earth or live in trees where they spin a web. These are found in the family known as Theraphosidae, about 700 species that are commonly referred to as tarantulas.

Many of these species are tottering on the edge of extinction because of habit destruction. This is why tarantula breeding is becoming increasingly important so it reduces the need to take popular species from the wild.

TARANTULA ANATOMY

The tarantulas possess the features already discussed for the phylum Arthropoda and the class Arachnida, so here we need only look at other aspects of their basic anatomy.

Hair: An obvious feature is that their legs and body are covered with hair. However, lots of other spiders also have hairy bodies, so this is by no means unique to the tarantulas, though it is prominent in these mygalomorphs. The hair, especially that of the abdomen, sheds readily and can prove rather irritable to human skin, which is why these spiders should be handled with care, especially by those who are aware they are sensitive to the hairs. On the underside of the leg tips the hairs may be iridescent.

Jaws: In mygalomorphs the jaws, unlike those of the "true spiders," open vertically to the body axis and strike downward into their prey, rather as do the fangs of a snake or cat. The mouth, situated just behind the fangs, is very small and simple, the food being crushed by small teeth at the bases of the pedi-palps and front legs. The spiders do not eat solid foods but inject a digestive enzyme into their prey that rapidly breaks down the tissues of internal organs into a liquid state. Mygalomorphs crush the exoskeleton of their prey (if it is an arthropod) in order to more rapidly pour these enzymes onto the soft inner parts.

Digestion: The liquid food is drawn into the esophagus by the pumping action of what is called the sucking stomach. This, using different muscles, then pumps the food into the intestines. Fecal matter is voided via the anus, the most dorsal of the abdominal openings.

Sex Organs: These are situated on the underside of the abdomen and exit via the genital opening. Males possess a number of testes while the female has a pair of ovaries. She also has a number of sper-matheca. These are small sacs, seen in many invertebrates, which are used to store sperm from the male in readiness for fertilizing her eggs. She can store sperm for many months in some species, giving her greater flexibility to bear offspring only when conditions are good.

Breathing Apparatus: Mygalomorphs breathe only via book lungs for they do not possess the tracheae seen in the true spiders. Book lungs are so-called because in the chamber that houses them the lamellae are stacked in the manner of pages in a book. They may be compared with the gills of fishes, but unlike the

If you need a glove to handle your tarantula, you probably shouldn't handle it to begin with. The soft abdomen of the spider is its weak spot and easily ruptures if the spider falls.

I. FRANCAIS

latter they are not exposed directly to the air but receive this via a short tube that is situated just anterior of the genital opening. The lamellar plates are hollow, and blood runs through them. Air oxygenates the blood, it is assumed, by diffusion alone because breathing movements have not been observed in spiders.

Book lungs are regarded as being primitive breathing mechanisms and are aided in the more advanced spiders by tracheae that enable these spiders to maintain a more rapid metabolic rate and thus move more quickly than do the mygalomorphs. This is a relative term,

however, because tarantulas can move very quickly when the need arises.

SPIDER SENSES

Spiders in general, and tarantulas in particular, do not have very good eyesight in spite of the fact that they have eight eyes. These are placed at the head end of the cephalothorax and the main ones are those situated in the middle of the front row of eyes. Their prime purpose seems to be in detecting various levels of light and dark, though it is quite possible that they may be able to form crude images when very close to an object—close meaning

This Mexican Red-knee, *Euathlus smithi*, is at home in the leaf litter. Book lungs under the abdomen allow it to survive with very little air, just enough to support its slow metabolism. Photo: R. D. Bartlett.

when they can just about touch it with their legs! (The jumping spiders of the family Salticidae have excellent eyesight, but even this extends to only a few centimeters, just enough for them to spot and jump on a passing insect.)

Spiders do not have the sense of hearing as we know it, but they can detect the slightest vibrations and air movements in their immediate environment, the vibrations being picked up by sensitive hairs found all over their body surfaces. These hairs are connected to nerve endings that transmit information to the nerve ganglia, which equate to the brain in the higher animals. This is situated in the lower middle part of the head (cephalothorax). If only one or two hairs are moved from their normal position this can be enough to elicit a response from the spider.

Another type of hair connected to the nervous system is found only on the legs. These are especially sensitive to air movements, such as those created by animals ranging from as small as a flying insect to larger currents created by, for example, a bird's wings. The degree of pressure movement in the air can be translated by the spider in order that it knows whether the creator is small enough to eat or large enough to be a potential threat.

Yet other sense receptors are found in the hard exoskeleton cuticle and on the appendages. These enable the spider to control the position of its body and legs—they tell it where each is in relation to the ground, to a potential prey, or to an opponent. The sense of touch is very important to tarantulas, so all of these various sensory hairs work together to provide the spider with a complete mental image of what it is confronted by or dealing with.

Spiders must be capable of distinguishing between various liquids, otherwise they would not know whether they are consuming a food or something that was poisonous! They possess no nose or, as far as can be ascertained, any taste buds comparable to those of higher animals. The hairs of the palps must therefore have some form of chemical receptors, as might the lining cells of the

esophagus. Via these, dangerous liquids or foods not enjoyed can be rejected. The pharynx of the mouth contains hairs that act as both sensors and filters for the rejection of unwanted liquids and debris. In one way or the other tarantulas are thus able to exist within their environment extremely well even though their senses are rather different from those of humans.

is deposited before it is transferred to the palps and thence to the female. The silk glands are found on the ventral rear part of the abdomen. Their external apertures, located just below the anus, are called spinnerets.

The silk is a protein liquid that is forced out of the spinnerets and solidifies as it emerges. It is quite elastic and also is capable of stretching considerably

M. GILROY

In most tarantulas the spinnerets are long and multi-segmented, yet tarantulas build only primitive webs.

SILK GLANDS

The silk produced by spiders is extremely strong in relation to its microscopic diameter. This is well known to most people because cobwebs are so common. However, its original role, and the one it is still used for in tarantulas, is for egg protection and for making the small web on which the sperm of a male

Facing Page: An enraged Indian Ornamental Black and White, *Poecilotheria regalis*, is a beautiful but frightening sight. Notice the sensory setae on all the legs and the pedipalps. Photo: M. Gilroy.

before it breaks. There are numerous types of silk produced by spiders, each for specific purposes. What appears as a single thread to the naked eye is in fact a collection of threads. The web of a spider is quite unique to its species and is a means of identifying a species—if you know enough about webs. Tarantulas display a range of webs but there is still much to learn about them.

The webs and egg sacs are sometimes impregnated with the hairs of the spider as a defense mechanism. This may be so where such hairs create irritation to an intruder, including animals sniffing or touching the web or egg sac. The irrita-

S. A. MINTON

A tiny but deadly Australian trapdoor tarantula, *Atrax* species, from New South Wales, Australia. Size and fuzziness are not indicators of potential danger in spiders—some of the largest and most hairy tarantulas are relatively harmless.

tion is created by tiny barbs on the ends of the hairs, some of which may also carry toxins.

HEAT REGULATION

Although most people think of animals as being either cold-blooded or warm-blooded, meaning their body temperature is much the same as that of the surrounding air or water (reptiles, fishes) or is regulated internally to remain at a constant temperature (mammals, birds), many creatures are not so easily classi-fied. The spiders are in this category. A tarantula is able to maintain its body temperature slightly above or below the ambient temperature both by the way it places its body in relation to a heat source (usually the sun) and possibly by some internal mechanism. However, for the purposes of caring for tarantulas it is best to regard them as being poikilother-mic (cold-blooded), as they will become rather inactive should it get too cold. They might even die if they are native to tropical climates.

TARANTULAS AS PETS

Now that you have some idea what tarantulas are all about, let us consider why you or anyone else would want to keep such animals as pets, and what are the problems you might encounter if you do not look after them correctly. A good way of answering your queries on these fascinating pets is to do so by posing the sort of questions the average person will ask when they see them for the first time. Once the initial staring and general funny remarks about them eating you have been made, people then become curious, and all of the things they do not know about spiders become a source of intrigue.

HOW DANGEROUS ARE THEY?

Most tarantulas are not at all dangerous compared to many other pets, providing you respect what they can do to you if you are careless. As far as is known, no one has ever died from the bite of a pet tarantula, which is not to say a large one could not create a very painful wound if it did sink its fangs into you. To do this it would normally need a reason—such as being badly handled or picked up too often. Some do not resent handling, but others regard it as an unpleasant intrusion on their own privacy. Some species are much more placid than others in much the same way that some breeds of dog are calm, while others may be quite the opposite. Even within the same species two tarantulas can display opposite personalities.

As far as is known there are no diseases of tarantulas that can be transferred from them to you or your other pets. In this sense they are safer that most other pets, which can on rare occasions transmit diseases to you if your hygiene standards are lower than they should be.

You may note that I said that most pet tarantulas are not dangerous. This is

A typical tarantula threat posture as shown by a Honduran Curly Hair, *Euathlus albopilosa*. This position is equivalent to a cat with its back arched and tail hair erect and means: Don't Touch!

R. BECHTER

because there are some tarantulas that are dangerous. The term tarantula is a rather loose one and it is applied to some of the funnelweb species of mygalomorphs that can prove toxic. However, these are not the ones normally imported and sold as pets. Further, there may well be one or more tarantula species yet to be discovered that may prove more dangerous than those we are already familiar with.

CAN THEY BE TRAINED?

No. Tarantulas are not intelligent like dogs, cats, parrots, or even rats. Intelligence is the product of having a good memory and being able to store and to some degree evaluate such memories. Spiders have only a limited ability to memorize things. In a sense every day, thus every happening, is a new one to them. What was experienced yesterday has limited value to them today.

Of course, they do possess natural instincts and do react to many stimuli, such as heat, cold, food, and so on. You might be able to condition them to certain actions. For example, if you tapped their housing three times and then dropped in a food item, they might associate the two things and respond by leaving their home to get the item. But then again, they might not if they did not feel especially hungry, so all in all training spiders is not something worth considering.

Facing Page: Mites are arachnids that are very similar to spiders in many respects. Notice the four pairs of legs in this water mite. Photo: D. Untergasser.

Tiny but very dangerous, this female South African black widow spider, *Latrodectus indistinctus*, shows off the red hourglass on the abdomen typical of the genus.

P. FREED

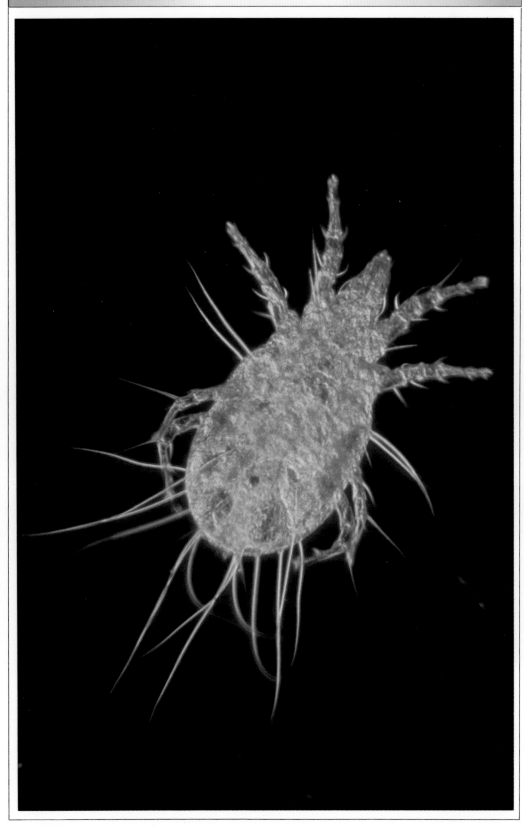

P. CARPENTER

This female Mexican Black Velvet, *Euathlus vagans*, can expect to live at least a dozen years in captivity. If you purchase a young captive-bred tarantula you can be assured of a long-lived pet.

DO THEY LIVE FOR VERY LONG?

It is generally accepted that the female of some species may live anything up to 20-30 years, the male considerably less— just two or three years after maturing. However, potential life expectancy and average expectancy in all animals are often totally different. A realistic expectancy for a pet female would be 6-14 years while that of a male may be from three to six years. Some species are notoriously difficult to keep alive for only a few years, probably because not enough is known about their husbandry needs.

Then again, much will depend upon the source of the tarantula. A wild-caught specimen may already be quite old, whereas with a captive-bred example you can be sure of its age. Further, the wild-caught example may be suffering from a parasitic infection or in other ways not be

very healthy, so might die as a result. A captive-bred spider should be free of infections and so should enjoy a healthy and long life.

CAN YOU KEEP TWO OR MORE TOGETHER?

Definitely not. While there are a few social spiders, tarantulas, with but one or two exceptions, are not among these. The only time these spiders come together is for mating purposes. Even that is a risky business for the male. If he does not go

Facing Page: The tarantula keeper should not ignore the pet potential of the large, colorful garden spiders such as this *Argiope*. Most, however, live only a season. Photo: M. Gilroy.

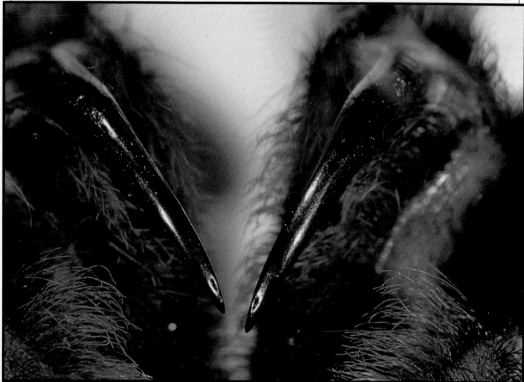

The fangs or chelicerae of a large tarantula, such as this *Theraphosa leblondi*, the Goliath Bird Eater, can inflict a painful mechanical injury. Notice the opening for the venom duct at the end of each fang.

about his courting with great care he could end up as a meal for his intended mate. Mating in spiders is not a social pleasure event, but a biological necessity for the continuation of the species.

Ordinarily, if two tarantulas meet they will fight and one will end up as the dinner for the other, or both will sustain very bad wounds and could die. It is therefore not cruel to keep a tarantula on its own because this is how it lives in the wild. Remember that it is not wise to keep two adult hamsters together, so there are other examples of animals that should be kept in a solitary state other than for breeding purposes.

DO TARANTULAS NEED LOTS OF SPACE?

Housing is discussed in a later section, so here it need only be said they do not. This makes them very easy to cater for.

ARE THESE SPIDERS EXPENSIVE TO MAINTAIN?

No. In fact, they will cost you less to keep in a healthy state than just about any other pet you could keep. For more on this topic read the chapter on feeding.

CAN YOU STROKE AND PET TARANTULAS?

Yes and no! Do bear in mind that these are not mammals, birds, or even reptiles, all of which can relate to their owner to a greater or lesser degree. Even a normally placid spider could become agitated by something and as a result bite you or throw hairs at you from its abdomen. In the latter instance the tarantula flicks its legs across its belly and the loose hairs are flung, often in little tufts, toward the perceived aggressor. These hairs can even get into your eyes. While not highly dangerous, the hairs do cause variable degrees of discomfort for a few hours or even days.

The actual bite of a tarantula may be compared with the sting of a bee, though possibly less painful depending on the species. It is dangerous only in that the wound could become the site of second-ary infection by bacteria. This being so, any such bite should be promptly cleaned and treated with an antiseptic.

If a tarantula is stroked it is wise to do this with the lie of the hair, thus reducing the risk of being "speared" by the hairs. However, never forget that you mean absolutely nothing to your spider in an emotional sense. Your finger is nothing more than a potential snack or a threat. The only reason that your pet normally will not bite is because it can no doubt detect from the sweat on your hand or body (which is made up of chemicals that the spider can analyze) that you are not an interesting dinner, so it has no reason to bite you.

You can certainly over-handle your pet, and when this is done it greatly increases the risk that the tarantula will show its displeasure by biting. In actual fact there is no reason at all why you should ever physically touch your pet if you do not want to. After all, you can gain tremen-dous pleasure from keeping fish without the need to touch them. The same is true of tarantulas. Many owners never directly handle their pets yet gain an enormous amount of satisfaction in just keeping and watching them.

HOW DO YOU HANDLE THESE SPIDERS?

Very carefully! There are two points you must always bear in mind when attempting to physically move a spider from one place to another. The first is that their abdomen is extremely delicate, thus vulnerable. The second is that you want to keep the risk of being bitten to a

This baboon spider, *Pterinochilus* species, means business. Baboon spiders are African tarantulas with a reputation for aggressive behavior and probably are not suitable for beginners.

R. BECHTER

R. REGAN

If you must pick up your tarantula, always hold the cephalothorax, never the abdomen. If you drop your tarantula once, you may never have a second chance.

minimum. With these realities in mind the safest way for you and your tarantula is if you usher it into a small container.

For example, if you place a small transparent plastic box over it and then carefully slide a flat piece of cardboard or plastic under the box the spider will walk onto this because it has no other choice—unless it walks up the container wall, when you can as easily place the 'floor' into position. You then can move your pet as required. Alternatively, you could coax the spider directly into a container and then place a ventilated lid onto this. This would be the safest way to transport your pet.

If you wish to actually hold your pet, then place your hand, palm upward, in front of it and gently prod the spider from the back with the index finger of your other hand or with a pencil or similar item. Your pet normally will sit quite still on your hand—but a nervous spider might just run straight up your arm! The problem then is that if it gets its feet caught in your shirt or whatever it will be much more difficult to remove, because it naturally will try to hang on like crazy.

If your pet sits still on your hand and you move around the room, a wise precaution would be to place your free hand just under the other hand. In this way if it decides to go for a walk you can place the free hand such that it steps onto this, then do likewise with the hand it was previously standing on. You can never be too cautious.

If you attempt to lift it bodily by placing

your fingers around its abdomen there is the chance that you might apply too much pressure and injure it or make it rather annoyed and more likely to try to defend itself. This could cause you to drop it, and tarantulas may easily burst their abdomen if they land badly. You would probably have a dead pet as a result because the open system blood circulation would quickly drain itself of blood. Spiders can in fact jump from quite a height and land safely, but a sudden fall is a different matter because they are not cats and do not always position their legs in readiness to cushion the fall.

Yet another method of lifting a tarantula is to gently but firmly take hold of its second and third legs near their bases on one side of its body, then lift. A spider is not accustomed to suddenly being whisked from terra firma and seems to be at a loss what to do, so it does nothing. However, if just one of its other feet are

More uncommon tarantulas, such as this Costa Rican Zebra, *Rhechosticta seemanni*, should be handled carefully, preferably by letting them walk into a glass or box rather than trying to pick them up with the fingers. Please do not do anything that will injure your pet. Photo: M. Gilroy.

R. D. BARTLETT

Beginners should be wary of purchasing, or at least of handling, unidentified tarantulas. Though there are very few records of serious accidents, there is no doubt that some tarantulas can cause serious bites in at least some especially susceptible humans. Large, smooth (not very hairy) tarantulas such as this African trapdoor spider deserve special caution. Your best bet is to never be bitten.

able to feel something solid before it is lifted it will wriggle and may attempt to bite your grasping fingers.

All in all, and especially if you are a little apprehensive of taking hold of your pet, it is far better to proceed using the safe methods first described. Never succumb to requests from friends to place your spider on your hand or body unless you are very confident to do so. Certainly you should not let other people hold your pet unless you know they are familiar with handling these animals. Otherwise they might suddenly become fearful if it moved. Their first reaction will be to drop it, with potentially fatal consequences to the spider. Finally, never use you pet to frighten people because this only reinforces their deep-rooted fear of these animals.

IS ONE SEX BETTER THAN ANOTHER AS A PET?

Not really. The main benefit of the female is that her life expectancy is much longer. When females have eggs they are more likely to be aggressive, but the same is true of a male that is carrying sperm and looking for a mate, so it works both ways.

HOW DO YOU TELL THE SEXES APART?

This isn't always easy unless you can carefully examine the tarantula using a powerful hand lens. Immature males look much the same as females, while mature specimens of either sex are broadly similar. Sex determination is discussed in more detail in the section on breeding. Generally, the wild-caught tarantulas you will see in pet shops are males because these are much easier to catch in the wild. For example, I live in New Mexico and during the summer it is not unusual to see one or more of these spiders slowly walking across the highway or across our garden. They are males looking for a female.

WHAT DOES A SPIDER DO?

This is actually a quite common question from those not familiar with will come out and move slowly around its environment when it starts to get hungry. The only time you will see it move at anything like high speed is when it pounces on its next meal or when it scurries away if it is frightened. After it has eaten it will normally preen its pedipalps and jaws. It will make webs as and when these are needed or if repairs are required. Beyond this, your pet will be quite happy resting for very long periods. By this mode of life it is able to go for long periods without a meal, because it is not a creature that burns up calories (thus food) by running around for no reason.

WHAT ARE THE BENEFITS OF OWNING THESE PETS?

Essentially, the most obvious reason you would want a tarantula, or a dog, a fish, a parrot, or a horse, is because you

R. BECHTER

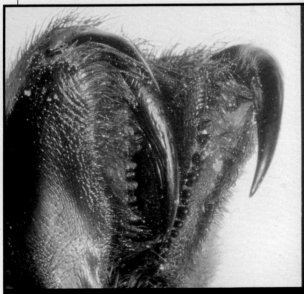

Views of the chelicerae of the Goliath Bird Eater, *Theraphosa leblondi*. Notice, in addition to the fangs, the numerous different types of bristles and their placement.

spiders. I am not quite sure what people think a tarantula should do, but the answer is, not a lot. Like any animal it does only what it needs to do in order to live. Your pet may sit in the same position for hours at a time. It may then retreat to its little home and sit there for days! It like them. It really is that simple. They fascinate you and you will probably find them interesting to study. Tarantulas come in a wide range of sizes and colors and these may be a source of esthetic beauty to their owner. Then again, many people like to breed them because this is

Owning an interesting tarantula such as this Indian Ornamental Black and White, *Poecilotheria regalis*, lets you become part of an alien world beyond the mundane items and events of everyday life. Tarantulas and other odd pets are a great learning experience and a great escape. Photo: M. Gilroy.

a challenging experience. The spiderlings can be sold at a profit, though few animal pursuits ever cover the total time and costs placed into them by their owners.

There still is much more to be learned about these pets and this may be another area of interest to the owner-student. The tarantula is not a costly pet to keep, nor does it take up much room. Further, when friends visit it is a sure-fire topic for conversation. You can exhibit your tarantulas and this provides yet another outlet for meeting people who share your passion for these fascinating pets.

There are many benefits to owning what is still a very exotic pet. I have even known people to keep them because they were arachnophobic, which means they were terrified of spiders. These pets helped them to overcome their fears of these and all other similar creepy crawl-ies. Often such fears stem from ignorance. Overcome the ignorance and the fear recedes or goes altogether and you become an arachnophile, a spider lover.

From keeping the actual spiders, some hobbyists then start to collect spider memorabilia such as spiders on stamps, spider T-shirts, mugs, pot spiders and many other items. Really, there are many benefits to studying these animals. The important thing is that you try to make it a two-way affair. Your prime concern should always be the welfare of your little pet.

Above: Watch out! This Mexican Red-knee, *Euathlus smithi*, is annoyed and ready to do something about it. Tarantulas don't have to bite to chase away predators—they can also kick some hair. A raised abdomen turned toward you is a warning to be prepared. Below: The fine white specks above the abdomen of the Mexican Red-knee are crystalline hairs kicked into the face of an aggressor. Many people are very sensitive to these hairs and may react with swollen eyes and breathing problems. Rule No. 1: Don't make any tarantula angry.

The bald spot on the abdomen of this Goliath Bird Eater, *Theraphosa leblondi*, indicates that the spider is a bad hair-kicker. A new coat of hairs will appear with the next molt.

R. BECHTER

HOUSING TARANTULAS

From the perspective of gaining the most enjoyment from your tarantulas and from their viewpoint, you should prepare the cage for your pet before you acquire the spider. It should mirror, as far as possible, the natural habitat in which the tarantula normally would live. This means you must do a little research and then obtain a species suited to the housing you have prepared. It would not be very wise to prepare housing suited to a desert ground-living tarantula and then place a tropical tree-dwelling species in it!

BASIC HOUSING

There is little need to discuss the materials needed for your terrarium (or tarantularium if you wish) because you cannot better a standard aquarium. These are made of plastic in various strengths or of glass.The latter is always the better choice because it does not scratch so easily, nor does it tend to yellow with age. The favored shape is a rectangle—tall for tree-dwelling species, low horizontal for burrow-dwellers. Avoid round aquariums for the simple reason that they can distort your view of your pet. There is no shortage of models from which to choose in your local pet shops, and those that are all glass or plastic are probably esthetically more pleasing than those with metal frames around them. However, you often can pick up good bargains if you read the pet columns of your local newspaper or visit yard sales. It is not important that your spider home is leak-proof, so an aquarist friend may have an aquarium he or she is discarding because it is leaking.

Tarantulas do not actually need a great deal of space in which to live, so a tank of say 25 x 20 x 20 cm (10 x 8 x 8 in), which is 2.5 US (2.2 UK) gallons, is adequate. Most keepers will prefer to use a larger aquarium of the following standard sizes: 45 x 38 x 30 cm (18 x 15 x 12 in) = 51 liters or 13.5 US (11.2 UK) gallons or 60 x

A simple but effective tarantula (and scorpion) terrarium. The light (red for evening viewing without disturbing your pet) and heating pad are controlled by a timer.

P. CARPENTER

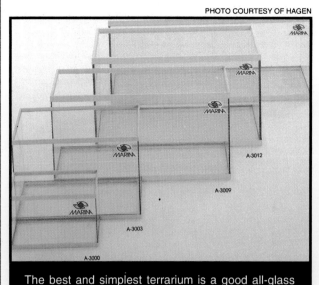

PHOTO COURTESY OF HAGEN

The best and simplest terrarium is a good all-glass aquarium.

a large aquarium you do have the option to later place a divider in it to form two units. However, if this is done be very sure the occupants on either side of the divider cannot possibly get over it, otherwise fighting will ensue. You can make a divider using a piece of glass cut to the needed size and then fix it in place with aquarium bonding silicone. Better still would be to fit four glass or plastic strips (two on each side to form channels for the glass) so the divider is not permanent and can be removed as needed.

SECURITY

It is very important that your tarantularium be escape-proof, otherwise you might just awake one morning to find that little Boris or Helga has decided to prove they enjoy a walk-about. Should this ever happen do not worry unduly, because such escapees have turned up weeks and even months after gaining their freedom. One day you might see them happily walking across your carpet from the secret domicile they have been living in. Ground-dwelling species will tend to look for a crevice at or just above the ground, while tree-dwellers will instinctively try to climb to a higher level. Bear this in mind when you commence your "find Boris or Helga"

30 x 30 cm (24 x 12 x 12 in) = 54 liters or 14 US (12 UK) gallons. These provide greater scope for you to landscape the terrarium even though your pet might not actually use this space on a day to day basis. However, in my experience, if a tarantula does have space it will from time to time utilize it when it goes on little hunting and exploratory expeditions. I have watched male tarantulas travel quite a way across my yard before disappearing into our horse pasture, so these spiders are not always as sedentary as some owners might suggest. In any case, part of the fun of owning these pets is that you want to try and make an interesting display that will fascinate your friends as well as be a source of pleasure to yourself. If the terrarium is very small then the comment that these pets do not do a lot will certainly apply because they will have no scope to do so!

If you purchase

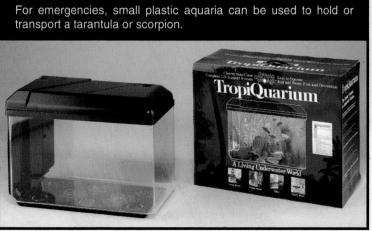

For emergencies, small plastic aquaria can be used to hold or transport a tarantula or scorpion.

PHOTO COURTESY OF HAGEN

PHOTO COURTESY OF FOUR PAWS

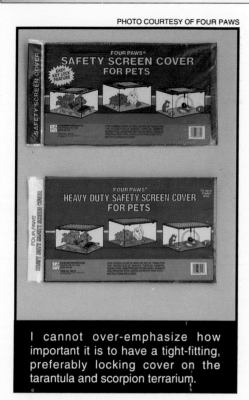

I cannot over-emphasize how important it is to have a tight-fitting, preferably locking cover on the tarantula and scorpion terrarium.

expedition. They will not go hungry because they will find enough little insects to munch on. Of course this would be a bad time to spray for roaches. In any case, these pets can survive for amazing lengths of time without food or water, but of course this is not good for them.

Aquaria often come with a hood (canopy) which, if it is a snug fit, will work fine for a tarantularium cover. The less expensive small plastic aquarium hoods could be dislodged by a determined pet, so it might be wise to place a heavy object on top to prevent this from happening. Larger aquaria may feature a hood and a glass cover. If the glass cover fits snugly it may provide a good safety barrier to prevent Boris leaving home, but be sure that if you have a desert tarantula it does not increase the humidity beyond bearable limits. Hoods also may incorporate one or more built-in lights that will provide some heat, especially if they are incandescent bulbs. Be sure that sealed hoods have a number of small holes drilled in them so there is ample air circulation.

An alternative to a hood is to fashion a neatly fitting piece of stiff wire gauze for the top. Be very sure it cannot be pushed upward, because these spiders really are amazingly deft at getting through an opening you would not think they could.

HEATING

Because your tarantula cannot regulate its body temperature as mammals and birds can, you must pay particular attention to this factor in the tarantularium. It should be maintained at about 70-75°F (21-24°C) for most species, though a few degrees higher may not be amiss and indeed may be essential for species from tropical zones. This said, I will add a comment or two. In the desert of the southwestern USA, which is home to numerous tarantulas, the overnight temperature from October through April, especially above 4,000 ft, can fall below zero. Further, the icy cold winds create an extra chilling subzero effect.

There is no doubt that spiders resident in such regions, in the comfort of their burrows, do not feel the freezing weather above them, but even so their ambient temperature will fall below that usually

To keep the terrarium just a few degrees above room temperature, a heating pad that fits under the cage is your best bet.

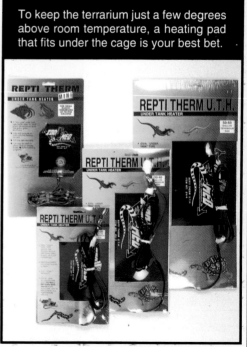

stated as needed by them. During the winter period some tarantulas, rather like goldfish and koi, enter a semihibernation period when their body metabolism is reduced to vital organ needs only. If they are kept at a constant temperature throughout the year this may affect their

you might need to experiment a little to establish how artificial seasonal temperature changes affect this or that species. Such changes must be gradual because sudden and erratic changes are most definitely not good for any animal if they occur over relatively short spans of time.

PHOTO COURTESY OF CREATIVE SURPRIZES

By attaching a colorful backdrop, you can create a unique appearance to the housing unit.

ultimate life expectancy, and it may influence breeding readiness and ability as well. In others words, many animals benefit from a short resting period at somewhat lower than usual temperatures and this is a natural part of their life.

Of course, at such times they will be very inactive—or not active at all, so you must balance your desires to see your pets up and about with their ideal biological needs. At this time in the hobby there is still a need to study this aspect of the lives of many tarantula species. This includes those from the tropics where it may well be very hot at lower altitudes, but where the situation changes markedly as the elevation increases, especially overnight. Seasonal temperature fluctuations are important to most animals, and matters such as the success of their molts are determined by fluctuations, so there is no reason to believe that tarantulas are any different. This would mean

As for the means of heating your terrarium, you have a number of options. These include space heating, heater pads, heated rocks, cable heating, terrarium heaters, and light bulb heating.

Space Heating: This is probably your best option because chances are your home is heated to the needed temperature for your own comfort. However, if you normally turn your heating off or low overnight it would be prudent to have a secondary system specifically for your tarantularium that is controlled by a thermostat and clicks on below a predetermined temperature —say once the room temperature falls to 65°F (18°C). This gives you the most economic heating combination. Bear in mind that if you keep two or more tarantulas and these are from differing parts of the globe, you may need a higher temperature for one tank than for the other. This is no problem if each tarantularium has its own heater.

Use a terrarium liner to give your pet sure footing when it is put in a temporary plastic terrarium during cleaning or feeding.

Heater Pad: These are available from pet stores and come in a number of sizes and heat output levels. They are placed immediately below the tarantularium. They are quite good for tree-dwellers, but not so good for burrowers. The latter will go deeper into their burrow if they get too warm—but then find that as they do so it starts to get hotter, which cannot be good for their general well being.

Cable Heating: This is a variation on the heater pad system and consists of special cables that can be covered with foil and then covered with planting soil. The tanks are placed onto this warm bed and provided steady heat that is thermostatically controlled. It is a system best used for multiple tanks in a line.

Terrarium Heaters: There are three types of terrarium heater that can be used for your tarantulas, with a fourth that you can fashion yourself. One is where a decorative plaster rock (a "hot rock") is built around a heating element so it has a dual function. These rocks come in different sizes to suit your needs. The second type is a small pad that is

Wind scorpions or solifugids like a few rocks in their terrarium but otherwise can be kept much like a scorpion or tarantula.

placed beneath a substrate of gravel and heats the surface area. Again, there are a range of heat output levels to suit your needs. The third heater is rather like a submersible aquarium heater, but is designed to operate in air without shattering. However, it would be wise to place metal gauze around it so there is no risk of your pet getting burned.

The fourth method you can fashion yourself in one of two ways. Firstly, you could place a low wattage aquarium heater in a jar of water and place gauze around the lid to ensure you pet does not fall into the jar! This method will provide low levels of heat—and it will add humidity to the air, which is important for

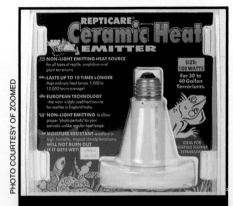

Black light heat emitters (which give off little or no visible light) can be used to heat a small terrarium.

Put your thermometer low on the side of the terrarium so it registers substrate temperature as well as air temperature.

tropical forest species. Alternatively, place a low wattage bulb under a tin can of appropriate size and place a shallow dish of water on this. The tin will heat up and in so doing will increase the evaporation rate of the water, thus again providing humidity. The tin provides a good area of heat, at the same time masking the light from the bulb, which might be too bright for your tarantula. You must take care that the tin can does not get so hot that it could burn your pet. A shield of gauze would again be useful.

Light Bulbs: These are best used outside of the housing unit so they can be raised or lowered to establish the required level of heat. The drawback to lights is that while they generate heat they also emit a lot of light, and this is not always appreciated by tarantulas. One way to

For tarantulas, a hot rock heater will work, but be sure there is a cool cage corner where the tarantula can burrow safely.

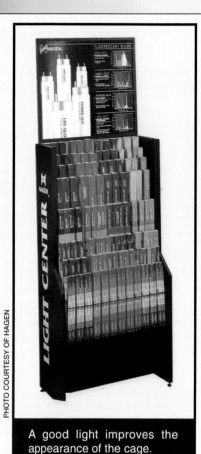

A good light improves the appearance of the cage.

overcome this is to use a low wattage spot that has a narrow diameter reflector on it. This can then be used to create a "hot spot" where the spider can bask if it so desires. It also creates an interesting esthetic look to the display. The major drawback to light heating is the amount of light itself, so I would recommend this method be avoided if possible: use a light for lighting and a heater for heating, avoiding the problems resulting when they are the same unit. Dull emitter infrared lamps are another good external means of heating, but whatever method you use it is very worthwhile having it wired through a thermostat so your spider will be neither too cold nor cooked!

You should include a thermometer in or on the tarantularium. In a large unit it is worthwhile placing the heating toward one end so that your pet can retreat to a somewhat cooler spot if it feels the need to do so. Be sure to test the heating system before your pet is placed into it.

In the event that you should experience a long power outage, it is useful to have some thick plastic foam or cork panels that can be placed around the aquarium. This will provide some insulation, while a hot water bottle beneath the tank will ensure that at least some heat is available for the tarantula.

WATER

Your pet does drink water, small though the quantity may be. A shallow dish will provide this, but you can also supply it by daily mist spraying any cobwebs or decorations. The spider will take its needs from these. Such water will be beneficial, indeed vital, to rainforest species that need higher humidity levels than those from desert regions.

If you use a lighted aquarium hood on the tarantula or scorpion terrarium, be sure that it fits closely on all sides and does not have cutouts for airlines and other fittings. Arachnids can be real escape artists.

DECORATION

The normal habitat of your tarantula will to a large degree dictate the sort of decoration best used in your terrarium. Arboreal species should have an upright log and maybe some imitation branches with leaves to give an interesting appearance and from which these species can spin their tube webs. Burrowing species should be provided with a substrate they can dig into. The best material for this is commercial potting soil because this will be parasite-free. Avoid using any natural materials such as garden soil, because these could contain parasites and bacteria that may be harmful to your pet.

The soil should be just moist, not a mud bath, and a light daily spraying with a mister bottle will keep it that way. It should have a depth of 5-15 cm (2-6 in). Dried cork and moss can be placed onto it here and there to add landscaping and to retain some moisture.

A piece of slate rock can be used at a slight upward angle in order to create a dark area to which your pet will no doubt retreat if the light is too intense for it. Be sure such slate is securely supported so there is no risk of it falling and flattening Boris! Rocks are not essential, but they do give esthetic appeal. If they are not too large your tarantula will clamber over them.

Avoid using sharp gravel as a part of

PHOTO COURTESY OF HAGEN

A few nice plastic plants will improve the appearance of your terrarium.

the tarantulascape because it might cause abrasions to the abdomen. If gravel is used at all it should be the smooth, small pebble type purchased from your pet shop. Rinse it in water before using it so it is clear of dust and debris. Dried cactus wood (cholla) makes a realistic part of the scene. If natural pieces of drift or other woods are used these should be soaked for 24 hours in a saline solution,

Backdrops offer many different scenes; compare the desert shown here to the mountainous panorama shown on page 31.

PHOTO COURTESY OF CREATIVE SURPRIZES

R. REGAN

A small plastic terrarium makes an excellent cage for a single tarantula. It is lightweight, easy to clean, and easy to securely close. A plastic terrarium is especially nice if you are raising several babies or keeping sexes separate before mating.

then thoroughly rinsed (removing all salt) and allowed to dry in the sun before being used. This will ensure they are free of parasites, though not necessarily free of their eggs. This is why imitation woods are safer and just as effective these days.

A ceramic plant pot half buried on its side in the substrate is another idea that might provide a nice little retreat for your pet. Some keepers believe that tarantulas are best kept in an almost clinical terrarium, but this always seems a very boring way to keep any pet. I think that when hobbyists read that a species is native to a desert region they have a mistaken idea of what such a terrain is like. It is actually full of various plants—cacti, acacias, mesquite, and many other shrubs—and there are always lots and lots of rocks. The earth surface is of a sandy clay-like consistency becoming more soil-like about 6 inches (15 cm) or so below this. This said, it is prudent not to include cacti in the limited confines of your tarantularium because of the risk that your pet might injure itself. Rainfor-

est species are, of course, to be found clambering over living and dead foliage, bark, and similar organic materials, as well as within the foliage of trees.

There is no valid reason, therefore, why you should not attempt to create a miniature ecosystem for your particular pet. In such a cage it will be far more likely to act as it would in its natural environment than if it is in a tiny aquarium that is totally unlike its natural home. Of course, you must try to keep out of this artificial home anything that could be injurious or likely to carry parasites.

CLEANING

Unlike pets such as fish, snakes, lizards, and amphibians, spiders are not animals that create a regular need to clean out their terraria. They may dirty the sides of their home because they will possibly attempt to climb the glass walls, and may have damp feet as they do so. They will probably attempt to dislodge any decorations if these are flimsy, and their burrowing habits will no doubt rearrange your idea of the perfect landscape. But as for actual cleaning needs, these will be limited to tidying up every month or so and keeping the glass panels clean. You will need to remove the skeletal remains of their food, and periodic changing or cleaning of rocks, wood, and the like would be sound husbandry.

The only specialized cleaning tool you might need would be an aquarium glass scrapper from you pet shop, and even this is not essential. Always make a habit of glancing at the thermometer each day just to check that the temperature is within tolerable limits.

PLACEMENT OF THE TERRARIUM

The final comment of note with respect to housing is that the completed tarantularium should be situated away from drafts, such as opposite doors, and on no account should it be placed where it will receive strong sunlight. Bear in mind that desert species stay hidden for most of the day and only begin their hunting forages as dusk approaches. Strong sunlight is not good for them.

With this fact in mind you might feature a blue or red nightlight in the terrarium or just outside of it, possibly controlled by a timer if your funds will allow it. By so doing you can turn out the room lights and watch what your little pet is up to as evening approaches. Blue and red lights seem to be invisible to most animals, tarantulas included, yet allow the human eye to observe movements easily.

If you must keep several tarantulas in one terrarium, use a partition to prevent accidents. Notice the water bowls with pieces of sponge in them to prevent drowning. Misting the terrarium and spider once a week might be safer.

I. FRANCAIS

FEEDING AND MOLTING

As neither of these two subjects will take up too much room to discuss, it is convenient to place them together. Further, as the molt is vital to a spider it is important that it is well fed before it enters each shedding of its old exoskeleton, so there is a link between the two subjects.

FEEDING

As pets go, your tarantula is about as easy to feed as you could wish. If it has a drawback it is that it eats livefoods, no commercial foods having been developed that these creatures can be fed on. If this would bother you, then I'm afraid a tarantula would not be a wise choice for you as a pet. However, their food consists mainly of other creatures that you certainly wouldn't want around your home, so for most people it is not some-

thing that causes them any problems. The usual diet of these pets will be insects, other spiders, and similar arthropods. The size of their prey will obviously be governed by their own size, so a spiderling would hardly be able to cope with a large insect, while a large tarantula would almost certainly ignore very small insects. Apart from arthropods, the larger "bird-eating" spiders will take pinky mice (mice just a day or two old), but this is not essential.

As for examples of your pet's diet, the following are but some of the more obvious: cockroaches, crickets, grasshoppers, mealworms, earthworms, maggots, flies, beetles, moths, aphids, butterflies.

Most owners will limit the diet to crickets and supply other prey purely as change of diet offerings on the basis that

Tarantulas and other arachnids partially digest their food outside the body, bathing it in digestive fluids from the mouth. The cricket being eaten by this Goliath Bird Eater, *Theraphosa leblondi*, soon will become an empty shell.

R. BECHTER

a little variety is the spice of life. Crickets and other suitable foods are bred commercially and sold in pet shops, largely for reptile owners, so you have no need to go on field trips. Indeed, it is better to purchase commercially bred insects and other foods because there will be much less risk of these harboring dangerous bacteria or parasites.

If you build up a collection of tarantulas you can of course keep overhead down by breeding your own insects from stock supplied by your pet shop, but if you have only one or maybe two tarantulas it is much easier just to purchase insects as and when you need them.

WHEN AND HOW MUCH TO FEED

Tarantulas are as variable with regard to their appetites as are people and most other animals. Some are greedy little critters, while others survive quite happily on what seem to be rather spartan rations. As a rough guide, a medium to large spider will consume three to five adult crickets per week. You can place one of these in each day and then see how your pet reacts when offered a couple more a few days later. Sometimes they will kill and devour two or three quite quickly, but at other times they may kill a few but only consume one or two. The others will probably be killed and wrapped in a silk cocoon for a later date.

A number of experiments have been conducted on these spiders and it has been established that they can go for many months without food and for over two months without water. I have read that one went for just over two years without eating, which is quite amazing. However, I certainly do not advise you to think in these terms, but rather to be sure your pet feeds each week, assuming it wants to. When they are about to molt they may refuse food, as they might at any time. For this reason it is best that you are satisfied they have been feeding well and are in excellent condition so they can easily go through a biological fast without problems. It is possible to tempt your pet to accept dead food by dangling it on a piece of string and jiggling this about in front of it. This ploy may work once or twice, but thereafter your pet may

well realize that the dead prey, or moist mincemeat, is not living and thus refuse the offering.

Perhaps surprisingly, some tarantulas will not always kill all the food prey that you offer them. For example, some cockroaches and beetles may be ignored after one has been killed. This may suggest that these spiders do have some notion of taste and find that certain creatures simply do not taste so good! The fact that they will subsequently leave these alone would also suggest that they can memorize certain things and recall such memories when making the decision not to attack this or that potential prey species.

Be sure water is available in a very shallow dish or by being sprayed daily onto their web or plants. A dehydrated tarantula will find it more difficult to move about its home because its legs require a certain liquid pressure in order to function correctly. They may be compared with the hydraulic arm of a mechanical digger that will fail to operate correctly if its fluid level is low. If water is lacking it will adversely affect the molting process and the spider's ability to hunt as quickly as it might otherwise. For these reasons never allow your pet to be without this readily available, but precious, liquid.

MOLTING

The exoskeleton of a spider is hard, so it is unable to expand in order to allow the spider to grow. To overcome this problem your spider will shed its old skeleton, allowing the new larger one to replace it. You may be wondering how the new "shell" can be larger than the old one if it has to fit inside the former. It is able to do so because it has an elastic consistency to it while it forms under the hard old skeleton. As the latter is shed the new skeleton pushes through and stretches before drying on contact with the air.

Initially, a layer of fluid separates the old and new skeletons, but this is ultimately absorbed into the new one and is replaced by a layer of air in order that the two skeletons do not become fused. The actual molting normally commences when a split appears on the side of the cara-

Feeding a tarantula in such a heavily decorated terrarium could result in a starved tarantula. Most tarantulas (and scorpions) are rather clumsy and could never find a cricket in this mess. If you like heavily decorated cages, feed your pet in a separate spartan tank. Photo: M. Gilroy.

hours in the case of a large mature tarantula. At such a time you should not handle your pet or in any other way disturb it. This could result in damage to its new and delicate skeleton. In the wild, spiders are very vulnerable during their molting process and often are eaten by birds and reptiles. Even some of their own prey species may attack them at this time.

Apart from this possibility, the actual molting process carries many inherent problems for a spider. If it is undernourished or dehydrated this may affect the chemical changes that take place as the new skeleton is formed, with many potential dangers. The skeleton may be too weak, or the spider may be too weak to shed its old skeleton.

Baby tarantulas undergo a number of frequent molts, maybe up to four a year, and in general cope very well with these, which are finished quickly and without problem. Once the spider is mature it will normally molt once, sometimes twice, a year until it reaches maximum size. Then the molts may be less frequent and are required mainly to replace worn tissue or missing appendages. In the latter case the replacement limb often is smaller than the other limbs but, after further molts, eventually attains the same size.

Returning to the molting process, once this has been done your pet will start to flex its legs and fangs. This is to ensure they become very mobile. If the legs are not flexed enough the spider will walk with a more stilted action until its next molt. This may be because of malnutrition or for any of many other possible reasons. Burrowing tarantulas often will spin a small web on the ground and then lie on their backs in order to molt, but this is by no means always the case. Some will find a suitable retreat and lean against this, while yet others may stay in an upright position and then lie on their sides when they actually push out of their old skeleton. Arboreal species normally will molt within the confines of their tube web.

Prior to the molt you may notice that the rear of the tarantula's abdomen is shedding hairs (see photos pages 26 & 27) and may turn a dark blue-black

pace part of the cephalothorax. This split continues horizontally along the abdomen and the old skeleton opens rather like a hinged lid. The spider then pushes its new body out first and withdraws its legs from their former skeleton. Once the spider has emerged it will rest for a length of time which can vary from just minutes with spiderlings to over 24

Tarantulas shed on a regular basis depending on their age and how much they are fed. The shed skin can be recovered and preserved dry in a frame on cotton for an interesting display. Photo: M. Gilroy.

P. FREED

Large wolf spiders, mostly members of the genus *Lycosa* and relatives, also make interesting pets. This female is carrying her entire brood on her body for their protection, a common habit among wolf spiders. An Italian wolf spider was the origin of the name tarantula for our big hairy pets.

color—though you will hardly notice this color change in a species that is already that color. Also, an elderly male may start to lose abdominal hairs, but such hairs may be missing if they have been used as a defense mechanism. Some spiders will refuse food for weeks before they molt but others will accept it almost up to molting time, so you cannot categorically say when your pet is about to molt. Likewise, after the molt some spiders will not feed for days or even weeks, while others will do so. Be sure you remove all food species from the terrarium once the molt is underway, and check that the temperature and humidity levels are satisfactory so that these do not become the source of molt problems. Make a note of the date of each molt.

Once the male has attained maturity, which many take two to five years depending on the species, he will molt only a few times before dying. While he is a juvenile you sometimes may be able to sex him on account of his more swollen pedipalps, but this is by no means a totally reliable guide. Once mature he will, in numerous species, sport small claw-like hooks (tibial spurs) on his first pair of legs, these being used during breeding. The female on reaching maturity will molt many times and live considerably longer than the male, more than twice his age assuming she remains in good health.

BREEDING

If you are about to purchase your first tarantula, it is unlikely that you will be thinking in terms of breeding it in the immediate future. Actually, it is best to gain some practical experience with these animals before you embark on what is a much more involved aspect of the hobby. What follows will give you an insight into the breeding of tarantulas so the information will be to hand if you need it at a later date.

GENERAL BREEDING DATA

The actual act of copulation in spiders is generally quite rapid, this being a desirable need in a species where the female just might eat her mate given half the chance! An average time would be no more than half a minute, though much longer pairings have been observed. Assuming the mating was successful, the female will develop the eggs over a period of five to ten weeks (depending on the species), at which time she will spin a silken web on which the tiny eggs will be laid.

In numbers, there can be anything from 50-700 eggs, again depending on the species. Where a large number are laid you can assume that in the wild the mortality rate will be very high, arising from a mixture of infertility, predation, diseased eggs, and cannibalism. Under captive conditions a much larger number of the offspring can be raised to maturity if husbandry techniques are adequate. This can prove a mixed blessing.

Once the eggs are laid the female will carefully wrap these up in silk so that a ball is created. In this the eggs will be incubated for one to four months. During this period the female will rarely stray far from her offspring. Once the eggs hatch the little spiderlings will quickly scuttle off to seek shelter. If they hang around too long their mother will commence dining upon them!

While the spiderlings are still in their silken cocoon they will hatch and experience their first molt. Once they leave the silk ball they will molt again about a week later and repeat the process as they grow.

For his part, the male tarantula on maturing will spin a special web on which he deposits his sperm. He will then alight on the web and, using his pedipalps, will draw the sperm into the special mating structures on the tarsal portion of the pedipalps. The tip of this special structure is called the embolus, and it is this that the male inserts into one of the genital openings of the female.

MATING

With general breeding facts now discussed we can return to the matter of the actual mating. Assuming you have a female, you will need to own a male or else be able to locate a hobbyist who has a tarantula of the same species and who is prepared to let you use their male, with its attendant risks, in return for a small fee or perhaps some of the youngsters that result. If you own a mature male that was costly to purchase you might have the mating conducted on the basis that there is an agreed fee, in whatever form this is to be, for a successful outcome, but also an agreement that if the female should kill your male then the owner must pay for a replacement. This way at least you are not taking the lion's share of the risk.

The male is always introduced into the terrarium of the female, never the other way around as is normal with most pet mammal species. It would be prudent to have a small stick and a piece of cardboard or stiff plastic available at this time. This is just in case you need to separate the intended lovers rather quickly. The courtship ritual within tarantulas varies from species to species, but in almost all instances the male goes about the task very cautiously.

He may pound the floor with his legs, as well as vibrate his body up and down. If the female is receptive to his overtures she will respond in a like manner, at which time he will advance toward her. At this point she will rear up and expose her fangs. His reaction is to rush in and use his tibial spurs to lock her fangs in the open position. Having neutralized her danger for the moment, he now proceeds to inseminate her with his sperm. This accomplished, he will release one of her fangs and pacify her by stroking her abdomen with his free front leg. Finally, the other fang is released and he will make a hasty retreat.

This is the hoped for way the mating will proceed, but not all matings go so smoothly, which is why the owner of the male is taking a risk in letting his or her pet be used for breeding. If the male is snack might be the order of the day, breeding matters having been attended to! Not all females of all species, or indeed individuals of the same species, are quite so aggressive, but it is perhaps better to assume the worst than have a fatality on your hands.

Once a mating has been effected you can attempt a repeat mating a few days later. It is generally held that if the female

R. BECHTER

The beginning of mating in the Mexican Black Velvet, *Euathlus vagans*, is seen when the male uses his first pair of legs to "smell" if the female is ready to mate. Often he busily is vibrating his pedipalps at this time as well.

sluggish locking his spurs onto her fangs, she may bite him, and that would be the end of his love life because she will then probably proceed to dine on him. This is much more likely if she happens to be rather hungry, in which case she may rush at him on sight. However, in the fight that may ensue it is always possible that she may be the loser, though usually it is the male. It is also possible that both may be badly injured.

If the female rushes at her intended mate, this is when your quick actions in bringing the cardboard between them may well prove a life saver. Likewise, once the mating has been effected you may need to act promptly in order to retrieve the male before the female decides that a rejects such a second mating it is a sure sign she was fertilized on the first occasion, but this may not always be the case.

REARING THE YOUNGSTERS

Once the eggs have hatched through the small aperture that appears in the egg sac, the female should be removed to a new home so that the youngsters can grow in safety. The baby spiders can be supplied with a pile of moss, and in this and the substrate they will live for a few weeks without problems. Of course, the larger and stronger ones will eat the weaklings, but this is nature's way and there is little that can be done to avoid the situation at this stage. Some fruitflies can be placed into the terrarium on a

R. BECHTER

The male Iridescent Pink-toe, *Avicularia metallica*, is trying to position the larger female (reared back and fangs exposed) so his tibial spurs will lock her fangs, allowing safe mating. This is a very dangerous move by the male, and sometimes he misses. A miss in this case can lead to death. Mating tarantulas is risky business.

regular basis to provide food for the growing spiders.

After a few weeks it will be time to separate the youngsters, otherwise the death rate will start to rise rather sharply. This is when the hard work really begins, because you may still have a hundred or more babies (maybe up to 400!), each now needing their own little house! A stock of small jars with gauze covered lids will be fine, but any sealed container you can obtain in quantity will be satisfactory. This should be half filled with potting soil and a little moss or cotton wool that will hold the needed moisture. The babies need to be fed at least twice a week, and with a couple of hundred or more youngsters this becomes a major chore. Microcrickets (pinheads) can also be part of the diet as the little spiders get larger with each molt.

CONTROLLED BREEDING AND SELLING YOUR SPIDERS

The fact that some species of tarantulas are prolific in their reproduction potential, coupled with the reality that these spiders must have their own housing from an early age, does present the breeder with problems on a large scale compared with breeders of most other pets. If you are not very careful you will soon have hundreds of spiders that you have no outlet for. It does not take long to saturate the local pet shop market and even the national market once a few breeders are having success with the more popular species. Further, the more you produce, the lower the market price will be for them. In just a few years you could find it simply was not worth breeding. Take these facts into consideration before you even think of breeding.

R. D. BARTLETT

There is decent money to be made by breeding and successfully raising tarantulas. Some species, such as this African baboon spider, *Pterinochilus* species, are actively sought as small captive-bred babies. Captive breeding assures a continuing supply of good quality pets without having to worry about environmental regulations and degradation.

If you really want to involve yourself with this side of the hobby, I would suggest a better plan of action would be to try and breed the rarer species and those that have proved difficult to establish. This way if success comes you will have a larger market at your disposal, and you will get better prices as well. But you will then need to move on to other species once the one you have starts to become more readily available. However, because new hobbyists want young adults, retain some stock because these will be more valuable once they are nearing maturity (but that could be a few years, so investing in spiders is akin to fine wine producers who have to store wines for years so they can mature).

TARANTULA SPECIES

Although there are known to be over 700 tarantula species, only a relatively small number is available to hobbyists. The cost can vary widely, depending on local availability, how rare the species is in the wild, local popularity, and whether or not the species is being bred in captivity. Some countries restrict or totally ban the export of their fauna, which clearly affects market prices. The price asked does not necessarily reflect the suitability of the species as a pet, because a readily available tarantula may not be one of the easiest to keep. If you are a first or second time spider owner I would advise you to keep only the more common and less costly species until you have gained a good working knowledge of their needs and have determined for sure if you really are an arachnophile. By taking this approach you are actually helping those species that may be in short supply in their wild state, as reflected in their rarity and high cost in the pet shop. These are best kept by the breeder who will attempt to reproduce them and thus reduce the need to take them from the wild.

The following listing includes most of the familiar species seen in pet shops, both cheaper and more expensive. The common names used are those most current in the English-language literature, while the scientific names (Latin names) are those that appear to currently be most widely accepted. Both the scientific names and common names of tarantulas have changed quite a bit in the last decade and probably will change again. In particular, the scientific names of several very common species have been changed. I've tried to list the older names where appropriate as well as the most current name.

EUATHLUS SMITHI: MEXICAN RED-KNEE
(See pages 8, 26 for photo)

(Formerly *Brachypelma smithi*. Often called Mexican Red-leg, incorrectly, Mexican False Red-leg would be better.)

Easily the most well known species kept as a pet, this spider is indigenous to Mexico and its southerly neighbors. So many of these tarantulas have been taken from the wild that the species is now protected under Article 2 of the CITES agreement (Convention for International Trade in Endangered Species). This means, in practical terms, that individuals sold in pet shops or by other sources have almost all been breed in captivity.

The popularity of the Red-knee is not difficult to understand. It is colorful, hairy, of good size, long-lived, and normally a quite docile and agreeable individual. The carapace is black, edged with orange. This latter color is found on the patella or knee (thus the name Red-knee). However, it also extends to the outermost tips of the tibia and to the pedipalps. The rest of the legs and body are black or shades of brown, becoming creamy at the extremities of the tarsus and metatarsus. The black is flecked with cream-orange of various intensities depending on the age of the individual, the ambient temperature, and the diet. Color in these and all other spiders is of course under the control of color genes, so selection for well marked and boldly colored individuals as breeding stock is the best policy.

The Red-knee, if it does show aggression, is rather fond of hair kicking, so always treat such an individual with great respect until you establish its nature. If you do not you might find a rash appearing on your arm. The wearing of clear safety goggles would be wise with such a pet just to ensure none of the hairs will get in your eyes during this getting to know you period. The Red-knee is a ground dweller and prefers a temperature in the range of 70-75°F (21-24°C). Maturity is reached by about five years of age, the female then living for a further ten or more years.

EUATHLUS VAGANS: MEXICAN BLACK VELVET TARANTULA
(Photo page 16)

(Formerly *Brachypelma vagans*. Also called the Mexican Red-rump Tarantula.)

A close relative of the Red-knee, this very attractive species is similar but lacks the orange-red and cream on the legs. The orange on the carapace is much more restricted to the edges only, but the orange flecking stands out more as a result. The

Black Velvet has a pleasing disposition, remembering of course that the nature of spiders often is unique to each individual. Its husbandry needs are the same as for the Red-knee, and it has proved a reliable breeder.

EUATHLUS EMILIA: MEXICAN TRUE RED-LEG
(Shown below)

(Formerly *Brachypelma emilia*. Also called the Mexican Painted Tarantula.)

Distributed throughout Mexico and into the Central American countries, this species is rather similar to the Red-knee. Its carapace is orange in the center, and the orange of the legs is more extensive. While its husbandry needs are the same, its disposition is not as friendly as the Red-knee, so handle this nice looking spider with considerably more caution.

EUATHLUS MESOMELAS: COSTA RICAN RED-LEG
(Not illustrated)

(Formerly *Brachypelma mesomelas*.)

Yet another species of this popular genus, *E. mesomelas* illustrates the problems related to common names discussed earlier in the text. Its basic coloration is much the same as for *smithi*, but it has a more mixed reputation with regard to establishment in the tarantularium. Some keepers have maintained and even bred it without undue problems. Others have struggled just to keep it alive. What you should bear in mind, if you are indeed sure you have this species, is that it is native to tropical regions where there is rainforest and thus higher humidity levels than in the drier prairie and desert regions where other popular members of the genus are normally to be found.

EUATHLUS ALBOPILOSA: HONDURAN CURLY HAIR
(Photo page 13)

(Formerly *Brachypelma albopilosa*.)

Native to Honduras, Nicaragua, and Costa Rica, this brown tarantula sports somewhat long curly pale orange hairs on its body and legs. While it may not win any prizes as the most striking of spiders, it does have a number of redeeming features. It is quietly attractive in appearance, has been bred without problems, has a healthy appetite, and has a most agreeable nature.

I must at once expand on the concept of an agreeable nature in a tarantula.

The Mexican True Red-leg Tarantula, *Euathlus emilia*, has become a replacement for the protected *Euathlus smithi* in recent years. A bit more aggressive than the Red-knee, it still makes an excellent pet and is bred in captivity.

P. CARPENTER

Unfortunately, not many tarantulas read books about themselves, so they are not aware they should be docile because the book says they are! More than a few hobbyists have been disappointed to find a so-called friendly species has turned out to be a real little demon. Rather more rarely, one finds that a species held to be aggressive turns out to be quite the gentleman or lady—if you have the nerve to test that out after having read they are really nasty critters!

With this species, as for all those native to rainforests, remember to maintain a high humidity level if you want your pet to remain in the best of health. The Honduran Curly (sometimes called the Costa Rican Curly) is a burrower. Males mature at a relatively young age—about two years old—so the female really is the better investment as she should be good for at least ten or more years.

GRAMMOSTOLA CALA: CHILEAN BEAUTIFUL TARANTULA
(Photo on page 1)

(Also called the Chilean Rose-haired Tarantula.)

This is one of two species in this genus that are well known to tarantulophiles. Its ground color is a rich milk chocolate brown that is amply flecked with cream-pink hairs. The carapace is a soft to obvious pink. The species is native to coastal rainforests. While its preferred temperature range is comparable to the popular Red-knee, a somewhat higher level of humidity will be appreciated. Like all the previous species it is a ground-dwelling burrower. Its nature is less reliable than the Red-knee, so it is suggested that you do not physically handle it or, if you do, only with great care and after you have ascertained just how it reacts to your hand when this is placed into its cage.

Looking rather similar, and equally as unpredictable, is the Chilean Fire Tarantula, G. spatulata. However, its carapace is a darker brown to black in a sort of starburst pattern. The femur part of its legs is a dark brown liberally flecked with creamy orange hairs, but unless you can make direct comparisons its colors are not so distinct from G. cala that you could

readily tell them apart. This species has bred well in captivity, where its needs are much the same as those of the Red-knee.

A third member of the genus that you might see for sale is G. pulchripes, the Pampas Tawny Red. Its ground color is darkish brown, but this is highlighted by the light brown "rings" that are seen at each junction of the leg segments. The carapace also is edged with a tan color, while the abdomen is flecked with tan-orange hairs giving the species a very attractive color-coordinated look. It displays a reliably docile nature, but I have not seen enough of these species to say whether this is normal for the species or just for a few gentle ones. Certainly other members of the genus are not as even-tempered, so maybe I have only seen the nice ones!

APHONOPELMA CHALCODES: MEXICAN BLOND TARANTULA
(Photo page 51)

(Also known as the Mexican Palomino Tarantula.)

This quite striking little tarantula is found in the southwestern USA as well as in Mexico. It is not as commonly available as some of those already discussed, but its numbers are on the increase in collections. The ground color is a light brown liberally flecked with creamy colored hairs that give it the blondish look indicated by its common name. The carapace exhibits a golden hue to accentuate the blond name tag. The tarsus part of the legs (those at the extremities) is a darker brown, sometimes black. The abdomen is a much darker color than the rest of the body and legs.

Although somewhat nervous in its behavior, the species is not aggressive, so it can be handled. However, there is a risk that it just might decide to take off and, as a result, injure itself if it fell from any height.

RHECHOSTICTA SEEMANNI: COSTA RICAN ZEBRA
(Photo page 21)

(Formerly Aphonopelma seemanni. Also called the Costa Rican Striped-knee.)

This is a very handsomely colored and patterned tarantula. The ground color is a

P. CARPENTER

Aphonopelma chalcodes, the Mexican Blond Tarantula, also commonly called the Mexican Palomino, is not a common pet and often is fairly expensive. The golden cephalothorax and legs are its major beauty points, and it is not a nervous species.

dark chocolate brown, almost appearing black at times. At each leg joint the hairs are banded in a rich yellow tan. The tarsus has a single stripe of tan along its length, this becoming a double stripe on the metatarsus and tibia before fading rapidly on the femur. The pedipalps also carry this double stripe, and the total effect is very distinctive. There are tan hairs on the body but insufficient to detract from the rich brown ground color.

The species has two minor drawbacks. It is rather nervous, so it will run rapidly if you place your hand or objects in its terrarium. But this has to be better than running at you! It is also rather destructive to any attempts to create a pleasing tarantulascape because it is an inveterate digger. The species is not the easiest to mate, but once this has been achieved they are reliable breeders.

Males mature relatively quickly at perhaps two to three years, and the female may live an average of about 12-15 years. The species inhabits rainforest areas from Costa Rica to southern Mexico, so it will appreciate humid conditions. I would suggest that you avoid handling this species not because it is aggressive but because, being nervous, it might easily fall from your hand or arm if it felt threatened.

AVICULARIA AVICULARIA: SOUTH AMERICAN PINK-TOE
(Photo on page 2)

Unlike the species so far described, this little gem is arboreal in its habits. It lives, among other places, in banana and pineapple plantations. As its scientific name suggests, this is one of the so-called "bird eating" spiders. Its ground color is a velvet black and its common name is derived from orange tips to the tarsal part of its legs. There are also some brown-orange hairs on other parts of its limbs. Being a tree-dweller, you must accommodate them in tall housing that features branches or bark to which they can anchor their extremely strong tubular webs. Their native habitat is tropical rainforest, so it is vital that humidity levels be higher than for desert species. Daily spraying of their web and the branches is easily the best way to provide their water needs. Much of their diet is winged insects. With this in mind you can supplement their normal cricket and grasshopper menu with a few moths. The males are somewhat smaller than the females. The disposition of the Pink-toe is fairly docile, but with a tendency to be rather nervous until it is very confident in its surroundings. It has the ability to jump, which might be disconcerting when first seen, but is part of their fascination.They are a highly regarded species that have bred well in captivity so should not be difficult to obtain. You may see other members of the genus for sale.

POECILOTHERIA REGALIS: INDIAN ORNA-MENTAL BLACK & WHITE
(Photos on pages 10 & 24)

(Also known as the Indian Ornate Rainforest Tarantula.)

This beautiful species is but one of a number of Asiatic spiders that are now becoming very desirable in tarantula circles. It is arboreal in its habitat so will need a tall tank and high humidity levels if it is to prosper under captive conditions. The Greek word *poikilos* means varied, many colored, or dappled and aptly suggests the pattern of the species.

The ground color is genetically what is known as brown agouti but much modified on parts of the body and legs. The agouti parts are very striking on the sides of the abdomen, where they are akin to the tabby pattern seen in cats. The hairs are a mixture of brown, yellow, and black, giving a grayish appearance. However, at intervals the black melanin pigment in the hairs is extensive and this creates the striped tabby pattern so well liked in felines. The back of the abdomen is white and this extends as a thin line along the cephalothorax. In both instances the white is bordered by black. The legs are irregularly banded in black and white, the front two pairs often carrying some yellow on them.

By nature this, together with related species of the same genus, is not the most docile of tarantulas so you are advised to handle it, if at all, with great care. It is also a very fast moving creature, which makes caution even more desirable. This apart it is an extremely colorful addition to any collection.

THERAPHOSA LEBLONDI: GOLIATH BIRD EATER
(Photos on pages 27, 38)

You could guess from its common name that this is a large spider and you would be correct. Not especially colorful, being shades of dark and light brown, it is desirable because of its size. Hailing from South America, it is a ground-dweller living in marshy swamp areas where it is known to burrow into the damp earth. Unlike many other tarantulas, the male of the species does not carry tibial spurs. Being a large spider it has a healthy appetite and this should be catered for by supplying not only larger insects, but two or three of them each day according to appetite. Never worry about over-feeding your pet because spiders will only consume what they need. As far as handling this tarantula goes the advice is "don't"! It is aggressive and can make an audible hissing sound when annoyed.

Because of both its unreliable temperament and its extra feeding needs, it is not really a suitable species for the first-time owner. It is discussed because you may see it for sale, so it is as well you know something about it. Of course, it will need a fairly spacious home. Its preferred housing temperature should be around the mid to upper seventies (Fahrenheit) with a humidity level of about 80%.

METRIOPELMA ZEBRATA: COSTA RICAN SUN TIGER ABDOMEN
(Photo on page 53)

(Also, much more simply, known as the Costa Rican Tiger-rump.)

Although it has a somewhat long and cumbersome common name, this is a small tarantula and is hardly cumbersome. It can move extremely fast when it wants to. Like the Goliath, the males do not possess tibial spurs, and like the Goliath it should be treated with respect even though it is hardly comparable in size. Its great appeal lies in the fact that it is nicely colored so is always a welcome addition to a collection if you get the opportunity to buy one. The ground color is brown, this being very dark and appearing almost black in parts. The carapace is a pink-orange of varying intensity, depending on the age and individual, while the abdomen is orange with a dark brown longitudinal stripe and some lesser stripes radiating from this.

Being native to tropical rainforests, be sure that the humidity level is up at around the 80% level otherwise your pet will suffer. It is a burrowing species, so the

R. BECHTER

There is not much doubt about the origin of the scientific and common names of *Metriopelma zebrata*, the Costa Rican Tiger-rump or Tiger Abdomen. Such a distinct pattern is not common in tarantulas.

substrate should be of moist potting soil.

OTHER SPECIES

The species discussed vary from docile to rather aggressive, and from plain to those that are very colorful and nicely patterned. Some, such as the Red-knees, you should have no difficulty obtaining, but others are less readily seen and likely to be expensive. From time to time you will see a number of the larger "bird eating" tarantulas for sale. However, I would suggest you consider carefully before these are purchased if you are a beginner. A number of them can be very aggressive. As always, there may be the exception to the rule. It is not always the nature of the spider that can be the problem. Some, being large and expensive, are relatively little-known in the hobby.

There is always the possibility that the spider might have some special needs that you are unaware of. Failure to provide these needs could mean the early death of your pet, which might be an expensive way for you to learn. It is much better to start with readily available species whose husbandry needs are well documented.

Sericopelma communis, one of many uncommon tarantulas sometimes found on the market. Photo: R. Bechter.

INTRODUCTION TO SCORPIONS

WHAT IS A SCORPION?

Unlike many pets, scorpions may be virtually unknown as wild animals to the majority of hobbyists. This group of the arachnids (order Scorpiones) is rather small by invertebrate standards (only 1000 to 1500 species, though many undescribed species are in the collections of the few experts on the group), is almost strictly nocturnal and secretive, and is

A small bark scorpion, *Centruroides vittatus*, from the southeastern United States. Notice the tail is curled to the side of the body, not over the top, a characteristic of the group.

mostly tropical in distribution. Though there are several dozen species in the dry regions of the western United States, only two species are common in most of the southeastern United States (both reaching north at least to Kentucky), and even Florida has only six or eight species. In Europe scorpions are few even in the Mediterranean countries (at least one is introduced into England), though they are abundant in species and individuals throughout Africa and Asia, as well as in tropical America and Australia. The average city resident in Europe and the eastern United States may never see a scorpion except in a pet shop.

In many ways scorpions are like drawn-out spiders. The body is of three basic parts: a cephalothorax or carapace that covers the head and the bases of the legs; a broad seven-segmented abdomen about the same length and shape as the carapace; and a five-segmented narrow "tail"

or postabdomen ending in a telson, also called the sting (not a true segment). The abdomen and the postabdomen together make up the opisthosoma, while the carapace often is scientifically known as the prosoma. In almost all scorpions there is a pair of large eyes about in the center of the carapace, and there are anywhere from zero to five pairs of smaller and inconspicuous eyes at the front corners of the carapace.

The telson of a scorpion is of great interest to hobbyists, of course. Usually it is somewhat bulbous in shape, with a pointed, claw-like sting at the top. Often there is a second small tubercle under the sting. Inside the telson are paired venom glands and associated muscles. As usual in venomous animals, the venom glands are under voluntary control by the animal. When the sting is tapped against either prey or predator, the muscles in the telson are constricted and push the venom glands against the walls of the telson, forcing out venom. The pedipalps and the chelicerae are never venomous in scorpions, though they can pinch. It is only the telson at the end of the tail that you have to watch out for.

As in the spiders, there are four pairs of true legs and a pair of leg-like pedipalps. In scorpions the pedipalps are large, usually held out in front of the living animal and end in large pincers, the chelae, that look like the claws of a lobster. These are the scorpion's major feeding appendages, used to catch and hold prey; they also are necessary to position the animals during mating, so a scorpion lacking the pedipalps is likely to starve and certainly cannot mate. (Scorpions do regenerate lost legs and even the pedipalps, if they survive long enough.) Between the pair of pedipalps at the very front of the carapace are the chelicerae, short, heavy pincers used for the final crushing and ripping of the prey. As the pieces of cricket or other prey are broken off, they are pushed back between the bases of the pedipalps, which have ridges and tubercles that help form the pieces into small particles that can be dissolved by enzymes emptied onto them by the mouth. Like spiders, scorpions eat only animal prey (especially insects, spiders, and small lizards and even snakes) that they can only absorb in predigested, liquid form.

GENERAL BEHAVIOR

Scorpions are nocturnal and secretive animals. If they have their choice, they are inactive during the day, emerging at sunset to feed and mate. The day is spent in shallow self-dug burrows in most species, though some types (bark scorpi-

ones, as you would imagine. Typically scorpions are some 2 to 4 inches in total length, but several species from the African and Asian tropics exceed 7 inches in length and may push 8 inches. At the other extreme are some truly tiny Caribbean and South American species of the genus *Microtityus* that are only a bit over half an inch long as adults. Size has absolutely no relationship to the potency of the venom, by the way, with small *Centruroides* species from the Americas among the most dangerous species and the gigantic *Pandinus* emperor scorpions of Africa being among the least offensive.

Few scorpions like the company of other scorpions, though they usually are

M. SMITH

The very plain Eastern Stripeless Scorpion, *Vejovis carolinensis*, is the other common southeastern United States scorpion. It is harmless and makes a fairly good pet in a humid terrarium at room temperature. Notice that the tail is held over the body, typical of most scorpions.

ons of the genus *Centruroides*, for instance) tend to rest above the ground, either hanging from under rocks and logs or under loose bark on low trees. Many species climb short distances into low bushes, and in the tropics it is not uncommon to find a few species in and near bromeliads and other plants growing on tree limbs. As a rule, scorpions are rather clumsy animals that tend to move by side-stepping or even backing up as often as they go forward. Never assume that they are sluggish, however, because all species can make short sprints fast enough to be on your hand or leg faster than you would like. The larger species tend to be more sluggish than the smaller

not especially cannibalistic. It is not uncommon to find large numbers of *Centruroides* and *Vejovis* under rocks and debris in the American Southwest, and specimens of about the same size usually do not see each other as prey. It also seems likely that most scorpions are relatively immune to the venom of their own species, much as are most venomous snakes. Bark scorpions (*Centruroides*) have been reported to form living ladders when large numbers are put into a bucket or collecting jar, literally climbing over each other until one reaches the rim, when many run up the "ladder" to escape. This is an important consideration to remember when setting up a cage for scorpions, especially small, agile species.

One last aspect of scorpion behavior or anatomy, depending on how you want to look at it, is that they glow in ultraviolet light. When it was discovered quite by accident in the early 1960's that alcohol in which scorpions had been preserved glowed when exposed to UV, it took scientists only a few years to utilize this property to find a way to collect large

legs. In most scorpions they look like a pair of short combs, with a solid strip at the front and numerous teeth at the back. Often male scorpions have more pecten teeth than females, but this varies tremendously. Between the two pectines is a rounded or somewhat triangular lobe, often divided, that is called the operculum or genital plate. The sex organs of the male are under the operculum, as is the

P. FREED

An unidentified scorpion, probably a *Tityus* species, from Costa Rica. Beware of scorpions with long, narrow claws (pedipalps)—many of these can be very dangerous or even deadly.

numbers of scorpions where previously only a few specimens had been taken. Today scorpion collectors go out on moonless nights with small portable UV (especially long-wave) lamps. Hundreds of specimens of formerly very rare species can sometimes be taken in a matter of hours, and it is possible to observe the natural activities of wild scorpions in some detail. (Caution: UV lights are harmful to human eyes and long exposures may cause blindness.) The use of UV lights for collecting scorpions has led to the discovery of numerous unknown species even in areas that seemingly had been well-collected.

PECTINES AND MATING

Scorpions have a pair of unique sensory organs, the pectines, originating between the bases of the third and fourth

opening of the female's reproductive system. When mating, the male scorpion grasps the pedipalps of the female with his pedipalps and the couple walk along "hand in hand," the pectines checking out the ground at regular intervals. Apparently the pectines are letting the scorpions check the nature of the ground to find a spot satisfactory for the next step of mating. When the right spot is found (often the top of a flat rock), the male drops a small stalked or cap-like droplet of sperm in a supporting jelly, the spermatophore, and then directs the female over it. She picks it up with her operculum, thus completing fertilization as far as the male is concerned.

Anywhere from two or three months to eight months later the female will have the abdomen conspicuously swollen, showing the membrane between the segments. All

All scorpions give live birth, a fascinating process to watch. Often pregnant females are imported and give birth within a few days or weeks of being purchased.

This emperor scorpion, *Pandinus imperator*, is producing young typical of all scorpions, white, helpless, and unable to sting or feed.

R. BECHTER

They will ride on their mother's back until their first molt about a week after birth.

scorpions give live birth, the number of young varying from just one or two to more commonly two or three dozen; a few species give birth to almost a hundred young. The young are born as tiny white duplicates of the mother. They are born one at a time, often still enclosed in thin egg membranes. In most American scorpions the female stands on the tips of her hind legs during birthing and forms a "basket" from the front legs to catch the young as they emerge from the operculum. For the first week or so the young (often called nymphs or larvae and also known as first instar stages) ride on their mother's back and do not feed, surviving on stored yolk or a similar food from the egg period they passed within their mother's body. After their first molt to the second instar stage, they still tend to ride on their mother's back but soon scatter and dig their own tiny burrows. The mother does not eat her young during their larval life but does defend her brood from predators. She also may provide tiny pieces of food for her young after their first molt, but by then she can make mistakes and occasionally eat a young that falls off her back. The larval scorpions, as mentioned, do not feed and thus do not eat each other. This helps make rearing young scorpions relatively simple; just leave it to momma for the first week or so.

Many of the smaller scorpions, at least the American and Mediterranean species, grow rather fast and may be sexually mature in just six or eight months. Other species, especially the larger ones of many different genera, seem to grow much more slowly, and it could be two to seven *years* before they reach full size and are ready to mate. In most scorpions mating usually occurs in the spring, with birth occurring during the hot summer months, but some species seem to be very non-specific and may mate and give birth at any time of the year that conditions are suitable. Males, incidentally, do not die after mating and seldom are attacked by their mates. In fact, a male may mate with many different females each year.

WHY KEEP A SCORPION?

I have to admit that scorpions are not exactly mainstream pets, and even today anyone who keeps scorpions is likely to be looked upon as slightly "teched in the head," as the old saying goes. Unlike tarantulas, which are generally harmless and placid pets, many scorpions, including some commonly available in the hobby, really are deadly. And I don't mean deadly in the theoretical or laboratory sense—they cause many deaths around the world each year, including mostly children but also some adults. Unless you know what animal you are keeping, you might as well be keeping a cobra.

This, unfortunately, is one of the major reasons many people keep scorpions today. They seem to enjoy the "thrill" of possibly being stung and seem to think they are immune to the effects of even deadly species. They like to show off their deadly pets to their friends and, like some snake keepers, may even brag about how often they've been stung. This is the wrong reason to keep a scorpion, but I suspect it is the driving reason for most keepers.

The real reason you should keep a scorpion is to observe the behavior of a fascinating group of unique animals that adapt well to captivity and can even be bred with little effort in some cases. Scorpions have a long fossil record (they once were several feet long, had gills, and lived in the ocean—or at least that is the way the fossils have traditionally been interpreted); they were among the first animals to emerge onto the land during the great transition from water to air. Today scorpions remain among the least understood animals though they may be abundant near major universities, and hobbyists who take the time to carefully observe their pets and keep records really can make significant additions to knowledge of the group. Besides, unlike tarantulas, scorpions actually do things in the terrarium. With care and the proper species, even the beginner can pick a nice, interesting, yet bizarre pet that may survive a dozen or more years in captivity.

SCORPION CARE & SELECTIONS

BASIC TERRARIA

Scorpions require very little in the way of high-tech housing. In fact, they can be kept in almost any secure container that can be kept at an appropriate temperature and humidity. Your best choice would be either an all-glass aquarium (only because metal-framed tanks are heavier and harder to handle) or one of the plastic terraria designed for keeping small mammals. You want a high terrarium, never a low one, because most scorpions have a tendency to climb. You also need a securely fitting top with at least some ventilation areas that either are screened with fine mesh or can be screened by you at home. If you plan on eventually keeping one of the more dangerous species of scorpions, do yourself and everyone else a favor and get a lid that will accept a padlock to prevent accidental opening of the cage.

You can keep most scorpions in a 5-gallon terrarium and they will be perfectly content. Even a couple of emperor scorpions will do well in such a small tank. However, if you want to keep four or five scorpions in a community tank (yes, most—but not all—scorpions can be kept in small groups and if given enough cover will not kill each other), start with a 20-gallon terrarium and prevent problems that may occur with overcrowding. The larger floor area will make heating easier.

Because scorpions, especially the smaller burrowers, can move very fast when they want to, terraria with vertical sliding doors are not recommended. You do not want to have the opening of the cage at a level that the scorpions can readily reach. Even with a secure top-opening lid, make sure there are no decorations or equipment closer that 8 or 10 inches from the lid. Aquarium tanks with glue seams at the joints present a possible problem because some scorpions, especially the bark scorpions, can climb these to the top. Caution always is advised. Don't take your pets casually, or you will get stung.

HEATING

Scorpions are nocturnal animals and they do not need to bask. They spend the day in burrows or under rocks and logs and do not do well in full light. The

Opisthophthalmus carinatus, a 4-inch desert scorpion from southern Africa. The dry savannahs and deserts of southern Africa are rich in scorpions.

P. FREED

popular concept of scorpions moving about on the surface of the desert during midday is wrong; scorpions do not balance their body temperature well in full light and can die in a few minutes if not allowed to burrow or seek shelter.

The best heating for scorpions is undertank heating, either a pad or heat strips. Good pads that maintain the temperature at 80°F (27°C) or so are readily available as they are used for lizards and snakes. Heating strips, cables, or coils are available at many pet shops and at nurseries and plant stores. They maintain a low and constant heat (also nearly 80°F) and have the advantage of being flexible. Thus they can be gently bent around the base of the cage or extended along the top of the table or terrarium stand under several small terraria. You want the soil temperature to be only a few degrees above average room temperatures and held at a constant temperature. If you want to try dark-light ceramic heaters over a flat rock in one corner, fine, but it's not necessary.

One problem with heating the cage is that you don't really want the entire bottom of the tank to be warm. Like any other animals, scorpions need a temperature gradient: warmer in one area of the cage, cooler in another. This is where the bottom area of the terrarium becomes important, as the larger the area the easier it is to create a gradient. You want to put the heating pad or strip under only a third or so of the tank, preferably to one side. This leaves the other side at room temperature or less and makes it easy for the scorpion to retreat to a cool burrow if it feels too warm. In a small cage, a pad may make the entire substrate too warm for the scorpion and it will not burrow, staying on the surface as far as possible from the heater. Strips and cables can provide a narrower area of heat that makes more of the cage bottom cooler potential burrowing areas for the scorpions.

Overheated scorpions stay on the surface, become very active at first, and appear uncoordinated. They may seem to sting themselves and appear to be "angry." Shortly they roll onto their backs and are obviously in great distress.

During summer hot spells you may have to reduce or turn off the heating entirely and provide deeper burrowing substrates so the scorpion can keep cool. A scorpion's burrow is a distinct microhabitat even in the desert, and it is your responsibility to allow the scorpion to duplicate it as well as possible in the terrarium.

SUBSTRATE AND DECORATIONS

Scorpions live in two basic types of habitats, and your choice of species will determine what type of substrate you must provide in the cage. Most of the larger scorpions, such as the species of *Pandinus* and *Heterometrus*, are from warm, humid habitats in Africa and Asia, so they need a loose, moist substrate for burrowing. The total substrate depth must be at least twice the length of the scorpion if your pet is to be comfortable. Start with 2 or 3 inches of a moisture-holding layer such as high quality peat moss or vermiculite and cover this with another few inches of orchid bark, small bark chips, or cypress mulch. You can use two or more types of covering substrate in the cage and see which the scorpion prefers. It might be best to make the substrate layer deepest in the coolest corner of the terrarium. Saturate the bottom layer (but be sure there is no open water—living scorpions are no longer aquatic as were their ancestors) when you first set up the terrarium. If you use peat moss, be sure to sterilize it first to avoid mites later. Vermiculite, perhaps the best substrate, should be rinsed to remove the small particles and then saturated but the extra water allowed to run off before the vermiculite is put in the terrarium. Spray the top substrate every day to two so it stays humid, but again, never wet. Beware of fungus and bacteria growing at the edge of the substrate and on old food remnants. A partially screened (use a very fine but tough gauze such as fiberglass screening) top should allow sufficient ventilation to prevent fungal problems. If you are using a plastic terrarium with the heavily ventilated lid usually sold with such small terraria, a sheet of plastic across most of the top of the cage under the lid should help conserve moisture and still allow access for misting and feeding. Some humid forest scorpions will drink from a very shallow water bowl.

Desert scorpions, such as most *Centruroides* and species of *Androctonus*, *Hadogenes*, *Opisthophthalmus*, *Buthus*, *Parabuthus*, and *Leiurus*, as well as most scorpions native to the American Southwest, need a dry terrarium with at least 5 or 6 inches of coarse sand for the substrate. The coarse calcium sand sold in pet shops is excellent. Fine building sand may cake and may contain harmful additives. Nothing else is needed for the substrate, just make sure it is deep enough to accommodate the scorpion. Desert scorpions need little water and should not be misted. If you want, provide a small, shallow water dish, but it probably won't be used.

Scorpions like to burrow under or at the edge of a rock or piece of wood, so these provide the best decorations. Breeding requires a smooth, flat surface for the spermatophore, such as a flat rock. Cork bark and broken pieces of flowerpot work well as decorations. Never use so many or such complicated decorations that you cannot easily see where your scorpion is burrowed in. Complicated pieces of driftwood, for instance, may allow a scorpion to hide where you won't see it until your hand or arm gets a most unpleasant surprise. Keep it simple. Remember that the decorations should always be low so the scorpion cannot get near the lid.

If you remove old food pieces as you find them, you should not have to clean out the terrarium more than once or twice a year. Scorpions are clean, easy to keep animals.

FEEDING

All scorpions are hunters, catching and eating insects of all types, spiders, and the occasional small lizard or snake. The staple food in captivity is crickets, sized to fit the size of the scorpion. Adult emperor scorpions often prefer grasshoppers to crickets and find them easier to catch. Large *Pandinus* may take two or three adult crickets each week, while bark scorpions may need only one half-grown cricket each week. Baby scorpions will take pinheads and somewhat larger crickets fed twice a week. Pregnant and cold scorpions do not feed—a subtle reminder to watch the temperature in the terrarium.

Mealworms also are eaten, and large scorpions may take an occasional pinky mouse if you want to provide some variety. Actually, scorpions do not seem to get bored with their cricket menu, and they do not need vitamin and calcium supplements. Large scorpions can be very clumsy feeders and may have trouble running down a large cricket or a grasshopper, so some keepers "trim" the jumping legs of the food before feeding. This is cruel to the food, however, and it might be best to just chill the food a bit before feeding. Of course, you should provide the food at night, during the scorpion's normal nocturnal activity cycle.

HANDLING

First, and last, **don't handle your scorpion.** Though the *Pandinus* and *Heterometrus* species are considered to be relatively harmless and seldom sting, the same cannot be said of many other species readily available in the hobby. *Androctonus* and some *Centruroides* species, as well as *Tityus* and *Leiurus*, have caused human deaths and there is no reason to believe your pet is harmless if it belongs to one of these genera. Be especially careful with unknown or doubtfully identified scorpions from Africa and Asia. If you must handle a scorpion, use a pair of long (12 inches) forceps and cover the tips with a layer of thin foam plastic to protect the scorpion. Pick a scorpion up by grasping it with the forceps just below the telson or sting, never near the base of the tail or by the legs. Scorpions can be extremely agile and their tails can move at truly incredible speed and strange angles. To move a scorpion from one cage to another, herd it into a transparent tumbler or some similar container that can be covered with a piece of cardboard during the move. Never let a scorpion out of your sight when you are handling or moving it.

Though impressive to look at, the emperor scorpions such as *Pandinus imperator* seldom sting and are considered relatively safe.

SELECTIONS

A beginning scorpion keeper should stick to species with heavy pedipalps. The reason for this is simple: most of the dangerously venomous scorpions belong to the family Buthidae, which contains the bark scorpions of the genera *Centruroides* and *Tityus* as well as *Androctonus* and *Leiurus*. One of the main characters of the family is that the pedipalps are slender, with narrow palms and long fingers. Scorpions of other, more harmless, families also may have slender pedipalps, but the beginner should not take any chances. Scorpion identification is practically impossible for the amateur, so you had best stick with what you know.

The emperor scorpions, especially the West African *Pandinus imperator*, are common and excellent pets. Many specimens of these deep brown to black scorpions exceed 6 inches in length, with short, weak-looking tails and large, blunt stings. The pedipalps are very large and covered with coarse tubercles. Most imports come from Ghana and vicinity, so they should be kept in warm humid terraria with a mulch substrate over vermiculite (even if your dealer is temporarily displaying them on plain sand). A few other (unidentified) *Pandinus* species appear in the hobby on occasion and really should not be trusted as they seem to be more aggressive than the common *P. imperator* and some species may be more dangerously venomous. The sting of *P. imperator* usually is described as being about like a bee sting, but the possibility of allergies should not be discounted.

Almost as large as *Pandinus* species are the Asian forest scorpions, *Heterometrus*. These scorpions look much like emperors at first glance and are closely related. They require similar terraria, warm and humid with mulchy substrates for burrowing. Unlike *Pandinus* species,

This female *Androctonus crassicauda*, one of the most deadly scorpions, is carrying about 80 newborns. Notice the pecten sticking out between the bases of the second and third legs.

the greatly inflated pedipalps of *Heterometrus* species are at least partially smooth and polished, with the tubercles restricted to lines near the fingers. Both *Pandinus* and *Heterometrus* can be distinguished from other large scorpions by looking at the front of the carapace, where there is a deep notch between the bases of the chelicerae in front of the large eyes (virtually straight in most other scorpions).

Bark scorpions, *Centruroides* and *Tityus* among others, include many bewildering species that come in a variety of patterns. They usually are under 3 inches long, have very long and slender tails held curled low and to the side (usually held high and over the back in more typical scorpions), and tend to have a small tooth or tubercle under the sting. The long, narrow pedipalps mean trouble, and some species (including *C. exilicauda*, an abundant species in Arizona) have caused human deaths, many human deaths in fact. However, the only really common scorpion of the southeastern United States, *C. vittatus*, makes a fairly

nice and safe pet and has only a mild sting (though take it from me, it can hurt, swell, and make you very uncomfortable for a few hours). Bark scorpions do not burrow like other scorpions, preferring to wedge their bodies into narrow spaces above the ground. They will hang upside down from rocks and logs and often are found several feet above the ground under the bark of trees and shrubs. It is very easy to have an accident with these scorpions, and they are great escape artists.

The American Southwest has a variety of harmless scorpions in the 2- to 4-inch range that make good pets though they seldom are available commercially. Unfortunately, they are almost impossible for the beginner to identify, which presents an element of danger in keeping them. The large (4 inches) hairy scorpions of the genus *Hadrurus* (don't be misled by the name, as the hairs can be restricted to the telson and the pedipalps and may not be very obvious) make good pets and sometimes are sold. They are considered to be harmless, though they have an

Androctonus bicolor, a deadly scorpion to be avoided even when available in pet shops. Photo: R. D. Bartlett.

uncomfortable sting. Keep all these species in desert terraria. Hairy scorpions, and probably most of the others, do not need any water and usually don't like to be misted.

Stay away from any *Androctonus* species that you see. There are many species, but as a general rule they can be recognized by the combination of narrow, long-fingered pedipalps and very broad tail segments that have high crests and rows of small tubercles. These are among the most dangerous of all venomous animals, and their venom has been equated with that of cobras and carpet vipers. They cause many deaths in northern Africa and the Middle East and, like most species with narrow pedipalps, are not suitable for the beginner.

IN ADDITION TO TARANTULAS AND SCORPIONS

If you are attracted to tarantulas and scorpions as pets, you probably would be interested in these related animals that sometimes are available at pet shops. All can be kept more or less like the basic pet scorpion, using a desert terrarium in most cases and feeding them on crickets and the occasional mealworm.

The exception to this care routine is the beautiful Golden Silk Spider, *Nephila clavipes*, and its allies of the southern United States and many tropical areas. Females have a body about an inch long (males are only 0.2 inches long), with a long golden yellow to honey brown abdomen with pairs of whitish spots. The first, second, and fourth legs have large tufts of "fur" that make identification simple. These spiders like warm, moist, shaded environments where they build their gigantic webs (up to 9 feet in width) in openings along trails and under trees. A part of the web is a bright yellow very sticky group of silk strands that have been noted to catch frogs and even small birds, though the more typical diet is various small insects. In the terrarium the web expands to fill the available space, as you would expect. These are beautiful, gentle animals that sometimes are sold in pet shops.

More often available lately are wind-scorpions or solifugids. In these usually desert animals the chelicerae are greatly enlarged and the most conspicuous feature of the head, while the first leg is tiny compared to the other legs and the pedipalps (which end bluntly in adhesive organs in some species) are large and leg-like. These animals like it hot and can be very active when they warm up sufficiently. Though they lack venom glands, they can catch and kill small lizards as well as the usual insects. Some quite large species (4 inches or more) from southern African deserts often are imported, but the typical American desert species are only an inch to two long. Unfortunately, these are short-lived animals, most only living a year. Though most are nocturnal, some species are active during the day, retreating to their burrows when the temperature gets too high.

The whipscorpions such as *Mastigoproctus*, the vinegaroons of much of warmer America, are 3-inch scorpion-like arachnids with gigantic pedipalp arms suited for crushing their prey and a thin, multisegmented "tail." Though fearsome in appearance, they are harmless. However, they have a ferocious pinch (and can draw blood) and often squirt from the abdomen a stinking chemical based on acetic acid. They make great pets if you can find one.

The same applies to the true tarantulas, the tailless whipscorpions of the genus *Tarantula* and their allies. These are greatly flattened arachnids that appear almost circular at first glance. The first legs are extremely elongated sensory organs and the pedipalps are large and covered with sharp spines and knife-like blades. These are common tropical animals often found under rocks and bark in humid areas (some like deserts, however) and once often imported with tropical plants such as bananas. They are common in southern Florida. Like other arachnids, they are nocturnal and feed on living insects. Some of the very large tropical species (6 inches or more across the legs) make spectacular and harmless pets.